The Glass Bathyscaphe

The Glass Bathyscaphe

ALAN MACFARLANE
AND
GERRY MARTIN

P
PROFILE BOOKS

First published in 2002 by
Profile Books Ltd
58A Hatton Garden
London EC1N 8LX
www.profilebooks.co.uk

1 3 5 7 9 10 8 6 4 2 1

Typeset in Fournier by
MacGuru
info@macguru.org.uk
Printed and bound in Great Britain by
Clays, Bungay, Suffolk

A CIP catalogue record for this book is available from the British Library.

ISBN 186 197 400 0

For Sarah and Hilda

Contents

Preface

THIS IS A BOOK about change, especially about how the presence of glass and the way in which humans have used glass has enormously accelerated change (and conversely, the absence of glass has slowed it down). There is a very strong human tendency when studying the past to try to identify individuals who have made history, to make them heroes, or at least key figures in explaining how events have unfolded. This is particularly tempting in trying to understand discovery and innovation. This tendency always lead to distorted history for change arises out of the combined activities of dozens, hundreds, or even thousands of individuals. Yet we often label a particular innovation or invention with the name of an individual as a shorthand or convenience. It is within that framework that the mention of particular named individuals in this book should be interpreted.

This also applies to the writing of the book. Though two named authors have written it it is impossible to disentangle their contributions. Likewise they are just part of a much larger network of friends, authorities and contacts who have contributed to this single book. Among those whose influence is most direct and obvious to us are the following. Professors Chris Bayly, Mark Elvin and Caroline Humphrey, and Drs Su Dalgleish, Simon Schaffer and David Sneath made various suggestions which were particularly helpful. Kim Prendergast carried out the survey of school classes and schoolteachers in South Korea and arranged our visit to a school there. Professor Tokoro, David Dugan and Carlo Massarrella helped specifically in relation to myopia in Japan and the latter two more generally in developing our ideas on glass. Our thanks also to Stephen Pollock-Hill of Nazeing Glass.

The whole book was carefully read and commented on by John Davey, Sally Dugan, Iris Macfarlane and Andrew Morgan. John Davey also acted as the editor. Mark Turin kindly checked the text in proof stage.

Sarah Harrison thought of the idea of narrowing down our focus to glass. She inspired us to put the academic paraphernalia and quotations elsewhere (www.alanmacfarlane.com/glass). She read through the text and made many valuable suggestions. To her, and to Hilda Martin who has also helped in many ways, we dedicate this book in gratitude for their support in our endless quest.

List of Illustrations and Credits

1

Invisible Glass

'... guessing before demonstrating! Do I need to remind you that this was how all important discoveries are made?'

Henri Poincaré

MOST OF US hardly give glass a thought, but imagine waking in a world where glass has been stripped away or uninvented. All glass utensils have vanished, including those now made of similar substances such as plastics which would not have existed without glass. All objects, technologies and ideas that owe their existence to glass have gone.

We feel for the alarm clock or watch: no clock or watch, however, for miniaturised clocks and watches cannot exist without the protective facing of glass. We grope for the light switch. But there can be no light switch, for there is no glass for the light bulb. When we draw back the curtains a blast of air strikes us through the glassless windows. If we suffer from short sight, we can see clearly for about ten inches. If we have long sight, as we probably do if we are over fifty, we will not be able to read. There are no contact lenses or spectacles to help us.

There is no clear mirror in the bathroom to shave by, no

bottles of ointments or glass for our toothbrush. There is no television in the living room, for with no screen it cannot exist. When we look out of the windows we see no cars, buses, trains or aeroplanes, for without windscreens none of them can operate (and they almost certainly have not been developed anyway). The shops in town have no window displays and our gardens no glasshouses. In the evening the streets flicker with torch-light. The central heating owes more to the Romans than the Victorians. We shiver in the darkness.

These are a few examples of what would be likely to happen if glass left our lives. Even more striking would be the way in which almost everything else would be affected. There would almost certainly be no electricity, since its first generation depended on gas or steam turbines, which required glass for their development. So there would be no radios, no computers, or email. There might well be no running water. Clearly we could not cook with electricity and there would be no freezers or fridges. There might also be surprisingly little use of non-human energy in what remained of industrial production. Our fields would produce less than one twentieth of their current yield without the fertilisers discovered by chemists using glass tools.

In our hospitals medicine would be killing more people than it cured. There would be no understanding of the world of bacteria and viruses, no antibiotics and no revolution in molecular biology from the discovery of DNA. As there would be little control of epidemic and endemic diseases these would everywhere be as rife as they were at the end of the eighteenth century.

Our understanding and control of space would be very limited. We might not even be able to prove that the earth goes round the sun. Our astronomy would be ancient and our weather prediction haphazard. Long-distance navigation would lack accurate tools for measuring longitude and latitude, and, of course, there would be no radar or radio communications, let alone the telephone and telegraph, to help us when we were lost.

The artistic and aesthetic world would also be entirely different. Not only would there be no photographs, films and television: our very concepts of space, perspective and reality would be radically different. There would have been no Renaissance discovery of how to represent three-dimensional space and our systems of representation might not be far removed from those of the twelfth century.

This book shows just how central glass is to every aspect of our lives. It is true that other substances, such as wood, bamboo, stone and clay, can provide shelter and storage. What is special about glass is that it combines these and many other practical uses with the ability to extend the most potent of our senses, sight, and the most formidable of human organs, the brain.

Through mirrors and lenses glass makes us feel differently about ourselves and about the world. Telescopes, microscopes and spectacles let us see the distant and the near in ways which the human eye unaided cannot do. Through barometers, thermometers, vacuum flasks, retorts and a whole panoply of other instruments, glass enables us to isolate chemicals and test theories about their properties and interactions. Glass allows a representation of nature to be captured accurately and stored and then transmitted over long distances without distortion. Glass, in short, influences every sphere of our lives.

Since the middle of the twentieth century, alternatives to glass have been developed and it now seems less irreplaceable. The windows of Chartres or King's College Chapel may never be exchanged for coloured perspex nor fine wine drunk from plastic, but glass is widely being superseded by other transparent materials. The time may come when our world would not collapse if glass disappeared as it undoubtedly would now, but this has no bearing on the degree to which glass technology was an important factor in improving wellbeing and knowledge in the millennia when modern civilisations evolved.

We live in a glass-soaked civilisation, but as for the bird in the Chinese proverb who finds it so difficult to discover air, the substance is almost invisible to us. To use a metaphor drawn from glass, it may be revealing for us to re-focus, to stop looking through glass, and let our eyes dwell on it for a moment to contemplate its wonder.

When we do notice glass we may find it difficult to place, for it tends to slip between categories. This is one source of its attraction and power. Glass is strange. Chemists find it defies their classifications. It is neither a true solid nor a true liquid and is often described as a 'fourth state of matter'. For a long time it baffled scientists, who could not find any crystalline structure within it. Glass is brittle, which is one of its weaknesses, but it is also enormously durable and flexible and, in the creative hands of an experienced and knowledgeable craftsman, it is almost infinitely malleable.

Glass, wrote Raymond McGrath and A. C. Frost in 1961,

> can take any colour and, though possessing no texture in the ordinary sense of the word, any surface treatment. As for responsiveness to light and shade, it has no serious competitor. It is capable of extreme finish and delicacy, is clean, durable and compact, and may be graduated almost imperceptibly from transparency through translucency to opacity, from perfect reflection through diffusion to the completely matt surface. There is, in fact, hardly any surface quality that it cannot assume. Yet at the same time it has a highly characteristic nature and in whatever manner we treat it or whatever surface we impose upon it, it still retains that unmistakable 'glassiness'. Whether it is embossed, engraved, painted, sand-blasted, mirrored, impressed with any pattern we choose, moulded, blown, flashed and so on – there is almost no limit to what it will endure or to the possible permutations and combinations of the different treatments – its vitreous qualities remain its decorative *raison d'être*.

In the early days of glass people were concerned less with its utility, which only later became apparent, than with its beauty. Glass was developed first to satisfy our aesthetic delight, later for its use in magic and then, through one of those great accidents of history, its light-bending capacities turned it into the most important avenue to truth about the natural world – a good illustration of John Keats's famous assertion that 'beauty is truth, truth beauty'. The awe-inspiring nature of glass was captured nearly sixty years ago by one of its great historians, W. B. Honey:

> Glass is nowadays too familiar to arouse all the wonder it deserves. Intrinsically wonderful as the product of mere sand and ashes it may be the occasion of further miracles when made into vessels. For its beauty never seems to be

> wholly the result of calculation. Its forms may be designed and controlled, its colour may be named and secured by a percentage of oxides; but beyond all these there is a quality in the material that defies prediction, and the play of light and colour within it, its insubstantial air, and the 'pattern of a gesture' which its form so often quite literally records, are only the chief elements, perhaps, in the beauty it may assume at the will of the artist.

The history of glass as a technology of thought has attracted surprisingly little sustained attention from scholars. Its development is commonly assumed to have been roughly similar over most of the world. If we think about glass at all, most of us assume that, having been invented some thousands of years ago, its making spread over Europe and Asia. It was then used everywhere in more or less the same ways and to the same extent, and that it has so continued up to the present. We may be dimly aware that it reached a peak in Venice during the Renaissance, but otherwise it mainly seems a very useful and available substance.

One purpose of this book is to give second thoughts to such assumptions and received wisdom. We want to share our surprise at discovering, for example, that glass was practically non-existent in most civilisations and that, where it was present, its role has varied enormously. We were equally surprised to find that it does not follow that once glass has been invented it will be used and also that some civilisations used glass and then gave it up. We hope also to recapture the sense of the astonishing nature of glass so vividly expressed by Dr Johnson in 1750:

> Who when he first saw the sand and ashes by a casual intenseness of heat melted into a metalline form, rugged

> with excrescences and clouded with impurities, would have imagined that in this shapeless lump lay concealed so many conveniences of life as would, in time, constitute a great part of the happiness of the world. Yet by some such fortuitous liquefaction was mankind taught to procure a body at once in a high degree solid and transparent; which might admit the light of the sun, and exclude the violence of the wind; which might extend the sight of the philosopher to new ranges of existence, and charm him at one time with the unbounded extent of material creation, and at another with the endless subordination of animal life; and, what is of yet more importance, might supply the decays of nature, and succour old age with subsidiary sight. Thus was the first artificer in glass employed, though without his knowledge or expectation. He was facilitating and prolonging the enjoyment of light, enlarging the avenues of science, and conferring the highest and most lasting pleasures; he was enabling the student to contemplate nature, and the beauty to behold herself.

To follow Dr Johnson's vision we travel widely in time and space, going back ten thousand years and moving over the whole known globe. The journey has not always been easy. To understand the enigmatic history of glass requires the insights and methods of many arts and sciences each of which has perceived a part of the story but, like the blind philosophers who each touched only one part of the elephant, cannot imagine the whole.

The lack of a rounded overview is well illustrated by the treatment of glass in the museums we examined when researching this book. The Victoria and Albert Museum in London and the Fitzwilliam Museum in Cambridge display fine drinking glasses and mirrors. The National Science Museum shows lenses and prisms and the British Museum archaeological and art objects. By assembling these collections in a virtual memory museum, we began to put together the shattered history of this extraordinary substance. But none dealt with windows. It was King's College Chapel, a few yards from where we wrote, with its medieval stained glass, that reminded us of one of the central roles of glass in history.

We found the fragments of an account scattered in the work of historians of art, technology and science, and of anthropologists, biologists, chemists and opthalmologists. Anyone who hopes to bring glass into focus thus has to travel lightly through many disciplines despite the warnings of good sense to remain within one's competence. We have thus been heavily dependent on experts in other fields, some of whose work is listed in the section of further reading at the end of the book. Because glass is such a complex substance and its influence so little studied, it can be difficult to prove its effects. We may sense, for example, that the mirror shaped our notions of the individual, or that lenses changed optics and profoundly affected the Renaissance. Yet it is hard to prove these connections beyond all argument. We suggest links and hope they are plausible and satisfying. People are wary of too much guesswork; this book contains a fair amount of it. However, we have not disguised our guesses when we have had to make them. It can also be fairly said that discovery sometimes occurs after a first rough set of guesses has begun to seem plausible enough to justify detailed examination. We hope our reasoning here will stimulate others to investigate the degree to which our arguments and conclusions are right or wrong.

During the last thousand years something quite extraordinary has happened in the world. The population of human beings has risen immensely, yet there is far more food to feed them as a result of changes in agriculture. The resources of available energy have vastly expanded. Life expectancy has generally increased as understanding of disease has improved. These and many other changes are part of the increase in reliable knowledge. That increase, we think, would not have been possible without glass. In telling something of its fascinating story we hope also to have shed some light on how our world came to be as it is and how we have come to be as we are.

2

Glass in the West – from Mesopotamia to Venice

Hail, holy Light, offspring of Heav'n's first-born,
Or of th' Eternal co-eternal beam
May I express thee unblamed? Since God is light,
And never but in unapproachèd light
Dwelt from eternity, dwelt then in thee,
Bright effluence of bright essence increate.

John Milton, *Paradise Lost*, Book III, lines 1–6

NO ONE IS CERTAIN WHERE, when or how glass originated, but for the purposes of this book this does not matter greatly. The where can broadly be answered by suggesting origins in the Middle East, perhaps in more than one place, including Egypt and Mesopotamia. As to when, some estimate the origins of glass at between 3000 and 2000 BC, while others suggest hints of glazing on pottery as early as 8000 BC. As to how, all is complete guesswork and we can only say that it was originally made by accident.

The earliest forms of glass were not transparent. We can see this when the number of recovered glass objects suddenly increases, in about 1500 BC. This was also the

1. Early Egyptian glass

An early Egyptian glass, made in about 1370 BC for storing liquid. Most early glass was opaque like this and it was well over a thousand years before clear glass became popular. Glass was thus long regarded as an alternative to pottery, or as a way of making replicas of opaque precious stones.

time when the 'core-formed' technique was developed. One of the two major techniques was as follows. A stick was covered in clay, then dipped into a crucible of heated glass and withdrawn so that it was covered in glass. The glass was smoothed with a slate and made to cover one end of the clay lump. After cooling, the stick was pulled out and the clay scraped away. Thus a hollow tube of glass, with whatever decorations had been scraped on the surface, could be made. At that point glass making extended along much of the eastern end of the Mediterranean and, through Phoenician merchants, was set to spread through the Greek islands and North Africa. It was seen as a substance which could mimic others, it was like clay, or could be used to imitate precious stones. It was not transparent. In this period glass was used for three purposes: to glaze pottery, for jewellery and to make small containers, mainly for liquids.

Somewhere between 1500 BC and the birth of Christ, perhaps around 500 BC, glass-making techniques spread to east Asia and were known to the Chinese. Thus by about 100 BC much of Eurasia had a common knowledge base of how to make coloured and plain glass. Its use continued to be mainly for the same purposes, glazing pottery, jewellery and containers.

Glass-blowing, which opened up endless new possibilities, was developed at some point in the century before the birth of Christ. Somewhere in Syria or Iraq, a revolutionary new technique of making glass artefacts was introduced. Up to this point glass objects were made by casting and grinding. Then came the invention of glass-blowing. This involved the use of a long iron tube of at least a metre in length which was dipped into the molten glass to pick up a lump. The blower then blew the glass into a bubble. We tend to think of this as an obvious development but it needed a highly sophisticated awareness of the properties of glass and its potential to reach this point. Technically it required heating glass to a much higher temperature than for moulding or casting. It needed to be very liquid. This required a knowledge and experience of furnaces that had developed in the glass industries of the Middle East. With the introduction of glass-blowing, really thin, transparent glass could be made. This new technique enormously increased the versatility of glass and, in particular, opened up potentially new uses.

It is really just over two thousand years ago that the great divergence in its use in east and west Eurasia began. In order to trace the causes and consequences we have to examine the various civilisations separately. We shall do so in relation to five major uses of glass. The first three are usefully distinguished using the more specific French terms for forms of glass instead of the collective English noun 'glass': 'verroterie' for glass beads, counters, toys and jewellery, 'verrerie' for glass vessels, vases, bottles and other utilitarian wares; and 'vitrail' or 'vitrage' for window glass. To this we may add two more. There are mirrors, and there are lenses and prisms, including such applications as spectacles.

The Romans have a central place in the history of glass. They provided not only the technical skills, but also a sense of glass as an important material in its own right. Roman glass technology was in many ways unrivalled until the nineteenth century. Yet it was the revolution in attitudes which was the most important feature in the development of the peculiar place of glass in the west. It is this attitude towards glass which distinguishes the history of glass in western Europe from its history in Asia.

The possibilities for innovation coincided with the peak of Roman civilisation, which placed glass at the centre of its interior decorative development. With the development of glass blowing it was possible to produce glass vessels cheaply and in large quantities. Glass was such a versatile, clean and beautiful substance that fine pieces became highly prized and symbols of wealth. Its success was so great that it began to undermine its main competitor, ceramics. Glass was principally used for containers of various kinds: dishes, bottles, jugs, cups, plates, spoons, even lamps and inkwells. It was also used for pavements, for coating walls, for forcing frames for seedlings, and even for drainpipes. It is no exaggeration to say that glass was used for a wider range of objects than at any other time in history, including the present. It was especially appreciated for the way it enhanced the attractiveness of the favourite Roman drink, wine.

In order to appreciate the colours of wine it was necessary to see through the glass. Thus another development, with great implications for the future, was the realisation that clear glass was both useful and beautiful. In all civilisations up to Rome, and in all other civilisations outside western Eurasia, glass was chiefly valued in its coloured and opaque forms, particularly as an imitation of precious stones. The long-term consequence of the perfecting of clear glass manufacture was the development of glass as a thinking tool, through mirrors, lenses and spectacles.

In their cutting, engraving, painting, gilt decoration and designs the Romans were greatly advanced. They knew all the tricks of the glass blower's trade and many of their fine pieces were as good as anything produced for many centuries after.

The technical ability of Roman glass craft workers meant that in terms of both the diversity of the objects which they made, and the quantity, it could be claimed that Roman civilisation was more glass-soaked than any other until the very recent past. This was partly due to the cheapness of the product. Every part of the huge empire could be supplied with glass. Rubbish heaps and middens suggest that glass utensils were thrown away when only slightly damaged as it was cheaper and easier to buy a new one than repair the old.

Of the five major uses of glass which we have suggested, the Romans developed two in particular, verroterie (glass beads, etc.) and verrerie (glass vessels and other domestic ware). But the other three uses, which would be the great developments of medieval Europe, although perfectly practicable and indeed known about, were not adopted to any great extent. These were vitrail (window glass), mirrors and lenses.

It is quite evident that the Romans could make good windows of glass, and occasionally did so. Window glass was apparently made by casting, and pieces of considerable size could be made. There is evidence for this from the Roman town of Pompeii. Other examples have been found in Roman houses in Italy and elsewhere. Yet most experts are surprised at the slow development of window glass in Italy. This is usually explained by the warm Mediterranean climate and the use of mica, alabaster and shells as cheaper alternatives. It may be that the large, flat panes of glass were rather crude and the imperfections meant that those who could afford them did not really see them as necessary for enhancing the beauty of their houses. For whatever reason, windows were not a major development in the south. As one moves towards northern Europe there is more evidence of glass windows, suggesting that the climate argument is correct. In Britain they were quite common after the Roman invasion and even reached beyond the frontier of the Roman Empire into southern Scotland. It was in northern Europe that window technology developed and flourished after the fall of Rome.

The Romans knew how to make glass mirrors, yet metal mirrors were preferred. Archaeological investigations have uncovered only a few examples of the former. The glass was generally coated with tin or, more rarely, silver. It was used in hand-mirrors but larger mirrors in which a man could see himself from head to toe have also been found.

Likewise, the use of glass to magnify objects was probably known. A little glass ball filled with water may have been used for fine work, such as engraving gems, but the Romans did not develop lenses, prisms and spectacles. Glass as a tool for obtaining reliable knowledge, either in optics or in chemistry, does not seem to have been developed to any significant degree. The Romans had laid the foundations for a world of glass, but the philosophical effects of this strange substance were not yet felt.

Our difficulty in understanding the influence of glass is partly caused by a widespread, but mistaken, impression. Most of us would assume that however wonderful the Roman glass was, all this was more or less lost to the west after the collapse of the Roman Empire. This makes superficial sense; the craftsmen would have been killed or dispersed and the market for glass would vanish. Furthermore, for a long period this assumption seemed to be borne out by the archaeological record. Far less glass dating after about AD 400 was dug up and what there was seemed to be of inferior quality. The apparent dearth of glass seemed a characteristic of Europe until as late as 1400; the destruction of Rome appeared to have left a glass vacuum for nearly a thousand years. Even experts on glass believed this until about two decades ago.

With such a picture it is very difficult to see that glass could be a technology which made a great difference. If western Europe more or less lost glass for a thousand years, it could hardly be in advance of other civilisations in this respect. And if there was little glass in western Europe by 1100, we would tend to think that it would take some centuries for it to revive, making its influence felt only from about the sixteenth century. This picture needs to be revised drastically. This will provide a very different link in the argument, for it can be shown that while there was certainly a decline in manufacture, much of the Roman legacy was maintained and, in some ways, improved upon even by 1200.

Archaeology can be misleading. Undoubtedly there was a very rapid loss in the quality and quantity of glass objects found after about the fifth century, and there seems to be little increase in the quantity even after the economic recovery of Europe from the eighth century. But this may not be a reflection of what happened, but rather of three other factors. Firstly, the development of Christianity meant that very few objects, including glass ones, were now placed in graves, the major source of Roman glass artefacts. Secondly, we know that broken glass was collected for recycling. Thirdly, much of the glass in Europe after the ninth or tenth century was made using potash made from the ashes of woodland plants such as bracken and beechwood, rather than from marine plants. The result is that glass objects made in this way are much more likely to decay than Roman glass, especially when buried in acid soils.

A final fact is that the assumption that glass had disappeared meant that those who did do some digging did not always notice what glass there was. This was compounded by the fact that until quite recently there was little serious archaeological work on early medieval Europe. The situation has now changed and the relatively young discipline of medieval archaeology has revealed a wealth of medieval glass. Excavations have transformed our knowledge and shown, as we might have guessed when we looked at the glorious stained glass in medieval churches, that it was widespread, made by craftsmen with great technical skill and confidence.

What then is the new picture that is emerging for the period between AD 500 and about 1200? It firstly incorporates a little of the old story, a half-truth which should not be entirely lost. It is indeed true that, particularly north of the Alps, the collapse of the Roman Empire led to a considerable loss of technical skill and quantity of output for a while. Yet rather than a complete break, we can now see a picture of a decline combined with continuity. There was a diminution of quality and quantity, but glass techniques were maintained and the high regard for the substance was not lost.

One of the reasons for our mistaken assumptions lies in the fact that much of the continued production occurred on the edges of the old Roman Empire. The centres of glass-making moved up into Germany, northern France and England. The Roman love of glass had spread as far as Afghanistan and the central Sahara, and as far north as Scotland and Scandinavia. When the barbarians overran the Roman Empire, they had already absorbed glass as an essential part of their lives, and this tradition continued. Thus recent research suggests that the Roman collapse had far less effect on the glass-makers north of the Alps than was once thought.

The Romans had learnt from the originators of glass in the eastern Mediterranean, but had then returned the gift by enriching glass-making there. Even before Rome collapsed, northern European glass-making was benefiting from the eastern influence and it continued to reinvigorate glass making after the collapse of the Roman Empire. A reservoir of skills and knowledge was maintained in Syria, Egypt and the Eastern Empire after the fall of the Roman Empire and it is quite clear that this had a dramatic effect on northern glass-makers. Immigrant glass-makers from the eastern Mediterranean spread across Europe and improved the techniques, particularly in north-western France.

Thus, after the fall of Rome the situation was far from static. Much of the old Roman technique was preserved, but through the centuries in the north the making of glass changed. For example, Roman techniques were replaced by new styles of glassware, mainly drinking vessels, known variously as Frankish, Merovingian or Teutonic. It is worth noting three especially important influences which bent the early Roman excellence in new directions as well as helping to preserve the great tradition.

One of these was Christianity and the introduction of glazed windows, particularly in churches, then the further development of painted and stained glass manufacture. There are references to such windows from fifth century France at Tours, and a little later from north-east England, in Sunderland, followed by developments at Monkwearmouth, and in the far north at Jarrow dating to the period between 682 and c. 870. By AD 1000 painted glass is mentioned quite frequently in church records, for example in those of the first Benedictine Monastery at Monte Cassino in 1066. It was the Benedictine order in particular that gave the impetus for window glass. It was they who saw the use of glass as a way of glorifying God through their involvement in its actual production in their monasteries, injecting huge amounts of skill and money into its development. The Benedictines were, in many ways, the transmitters of the great Roman legacy. The particular emphasis on window glass would lead into one of the most powerful forces behind the extraordinary explosion of glass manufacture from the twelfth century.

Apart from the addition of window glass, until 1100 the two main uses for glass had been for verroterie (beads, toys, jewellery) and verrerie (vessels). The period between 1100 and 1700 saw the continued development of the two earlier uses, especially in relation to fine drinking glasses. Windows, both of stained glass in religious buildings and of plain window glass for ordinary homes, became increasingly common. Similarly, there were improvements in both quality and size of glass mirrors as luxury items for the home. The new use for glass was in the development of lenses, prisms and spectacles, that is glass for optical purposes. Windows, mirrors and optical glass would change the knowledge base of Europe. None of these were manufactured to any extent elsewhere in this period.

It seems likely that the tradition of glass-making never died out in Italy after the fall of Rome, especially in the northern Adriatic area around Venice. Yet it was really from the thirteenth century that Italian, and particularly Venetian, glass-making began to influence the whole of Europe. By the early fourteenth century glass production was widespread and in the following century the techniques improved further, probably heavily influenced by events in the eastern Mediterranean. In particular the destruction of the city of Damascus (a great glass centre) by the Mongolian war-lord Timur the Great in 1400 probably led to an influx of craftsmen to Italy, as may also have happened in 1453 when Constantinople finally fell to the Turks.

Two particular technical developments in glass making laid the foundation for glass of a high enough quality to underpin the Knowledge Revolution. The first of them also shows how the recovery of the skills of ancient Roman glass-makers was an important influence. The glass-makers on the island of Murano, near Venice, experimented with Roman glass techniques and towards the end of the fifteenth century they developed a method of making glass in which thin canes of multicoloured glass are embedded, known as *millefiori*. Even more important was the development of crystal or (cristallo glass), a word first mentioned in 1409. It can be thin, almost weightless, free from flaws and colourless and it enabled the glass-makers to make wonderfully elegant and intricate forms. Its purity and thinness was an object of fascination and desire. Thus it fed into the artistic renaissance that was occurring in northern Italy at this time.

There is circularity in the development of glass in any civilisation that depends mainly on how it is perceived, but likewise, how it is perceived is dependent on its quality and versatility. As glass-making improved, so did the desire for it and hence money flowed into further improvements. Thus the glass-making explosion in Italy is not merely an automatic result of the increasing wealth of Europe from the twelfth century onwards. It is linked to many other forces, intellectual and cultural.

One of these was the growing fascination with curious and precious substances, particularly among Renaissance patrons. Rock crystal was especially prized, since it was thought to have magical properties. But it was available only to the rich. Fine glass became a cheaper substitute, just as beautiful and more versatile. The Venetian glass-makers also began to imitate many of the other beautiful hardstones such as agate, jade, jasper and lapis lazuli and to use them in all sorts of forms from cups to candlesticks as well as for beads.

The astounding versatility of the glass-makers of Murano was described by Georgius Agricola in his account of a visit in 1550. 'Glassmen make a variety of objects: cups, phials, pitchers, globular bottles, dishes, saucers, mirrors, animals, tree, ships. Of so many fine and wonderful objects I should take long to tell. I have seen such at Venice, and especially at the Feast of the Ascension when they were on sale at Murano, where are the most famous of all glass factories.' Glass making had become an important art form, an intellectual and cultural fashion, and this fed back into scientific and artistic experiments. Rich men began to set up glasshouses for their own use, out of curiosity and a desire to produce beautiful things. Glass-making became a noble pursuit.

Much stress has been laid here on Venetian glass, but it should be remembered that there were other important glass-making centres in Italy, particularly the northern town of Altare. Although smaller than Murano, the Altare works were particularly influential because it was their policy to spread their techniques as widely as possible, rather than to retain them as trade secrets as the Muranese attempted. Nor was glass making restricted to these two famous centres. Many other Italian cities, including Padua, Mantua, Ferrara, Ravenna and Bologna, had glass factories.

The influence of Italian glass techniques, as well as the glass itself, spread out all over western Europe, particularly from the sixteenth century. One important centre to which the new skills were transmitted was the Netherlands, and it seems more than a coincidence that one of the major northern centres of fine glass should be the other famous centre of Renaissance painting. Crystalline glass making reached Antwerp in 1537, and in 1541 a Venetian founded a mirror factory there.

Yet, although the Italian, and particularly Venetian, developments are of enormous importance after about 1400, they tend to distort the picture, especially for the period 1100–1400. Glass making was well developed in Germany and France at the end of the Roman Empire and this tradition continued, finding its highest development in Bohemia. Recent developments in medieval archaeology have now allowed us to see that fine glass was not an Italian preserve. There were, in fact, two different glass-making traditions in Europe. That in the north, in Germany, France, Flanders, Britain and Bohemia, was, certainly until the early fifteenth century, just as sophisticated as that in Italy, even if it used different techniques and produced other styles of glassware.

In Bohemia, the wealth created by the silver mines brought the prosperity that enabled people to buy the exceptionally fine, colourless and thin glass, which was being made there by the middle of the fourteenth century. This was the continuation of an earlier tradition and in due course the Bohemians would even outdo the Italians.

The development of glass did not come to a halt with the superb glass of Venice in the fifteenth century. Nor did glass-making remain basically an Italian and German activity. Its history from the sixteenth century was the gradual movement north until, by the end of the seventeenth century, the most advanced glass-making area in the world was England. From being a relatively backward area, England benefited from the influx of skilled refugees from the Catholic Counter-Reformation on the Continent. Thus, improved techniques and knowledge were fed into English glass-works. Glass-making on any scale needed enormous quantities of fuel to fire furnaces. From the early seventeenth century the shortage of wood led to another important development – the use of coal in English glass furnaces. This produced higher temperatures and lowered the cost of production of glass. These developments led into what was to be England's greatest contribution to the art of glass manufacture. This was the remarkable lead glass, patented by George Ravenscroft in the late seventeenth century, made from potash, lead oxide and calcined flints. It was to rival Venetian glass and could be mass-produced. It also had different light-bending properties from Venetian glass, but when used in combination with it enabled the development of powerful telescopes in the eighteenth century. The glass industry grew apace. In 1696 Houghton listed 88 glass factories making: bottles (39), looking-glass plates (2), crown and plate-glass (5), window glass (15), flint and ordinary glass (27). Twenty-six of these were in or near London.

The industrialisation of glass production in England, particularly through the use of coal of which there was

2. *An eighteenth-century English lead glass*
Developed in England in the last quarter of the seventeenth century, lead glass, because of its toughness and brilliance, became the most advanced form of glass in Europe. By an unexpected accident, this lead glass also eventually permitted the development of greatly improved telescopes and microscopes.

seemingly limitless supply, gave it an advantage at a time when there was an increasing desire for glass artefacts of all kinds. This in turn led to further innovation. The pattern repeats itself again and again with the push and pull of those skilled in glass-making seeking safe havens or new markets.

A civilisation in which glass had become ubiquitous, not merely for jewellery and utensils, but for mirrors, windows and lenses, had emerged. It was based on Roman foundations of knowledge and craft and this, combined with the rapid development of wealth in medieval Europe, meant a very steep trajectory or curve of knowledge and use. Between 1100 and 1600 a civilisation which used glass only in limited ways and in relatively small quantities had turned into one where superb glass was widely available. Glass had turned from a substance seen as a substitute for precious stones or ceramics to something entirely new. It provided accurate reflections of the observing individual, kept out the cold yet allowed the individual to see out of a building and helped humans to see the tiny and the close at hand in new ways. Both the speed at which this happened and the fact that it was such an extraordinary substance, altering human beings' most important sense, eyesight, had immense consequences. It happened at the right time, in the right places and with sufficient momentum to be the factor we are looking for. We now need to examine a little more closely how it worked its magic in science and art.

3

Glass and the Origin of Early Science

The One remains, the many change and pass;
Heaven's light forever shines, Earth's shadows fly;
Life, like a dome of many-coloured glass,
Stains the white radiance of Eternity.

Adonais, Percy Bysshe Shelley

THE ROLE OF GLASS in preparing the philosophical and practical foundations for the expansion of reliable knowledge from the time of Francis Bacon and Galileo at the end of the sixteenth century is largely invisible. Yet we need to understand something of what happened during earlier periods for at least two reasons. We cannot understand the burst of scientific activity from about 1600 without seeing that it is really a later wave in the growth of reliable knowledge in the west, which followed earlier waves, and in particular that which had started in the thirteenth century. Furthermore, it is deeply linked to developments in what we now classify as the 'arts'. Science and art, the pursuit of truth and beauty, were not separate endeavours. What we now call the 'artistic' Renaissance is also comprehensible only if we see it partly as an application of the discoveries in medieval geometry and optics.

By the time of Bacon and Galileo, there was already a four-fold foundation for science, without which the seventeenth century's developments could not have occurred. There was a set of techniques, what we call the experimental method. There was a certain attitude of curiosity, a belief in the possibility of finding new things, a confidence that there were deeper laws to be discovered behind the surface of reality and that it was man's task to discover these. There was a set of mathematical tools, particularly geometry and algebra, and a large accumulated knowledge of the natural world and how it worked. Finally there was already the concept of the laboratory, filled with tools for thought, many of them made of glass, but also others, such as the astrolabe, for investigating and measuring nature with precision. Considerably influenced by alchemical experiments there was an array of retorts, flasks, jugs, mirrors, lenses and prisms, already being used in chemistry, physics and optics. The emergence of all of this is not inevitable. Indeed, it could be said to go against some of the most powerful tendencies which we find in relation to human knowledge.

Einstein characterised science as the combination of two things: Greek geometry and the experimental method, and the latter he linked to the 'Renaissance'. Yet a great deal of research in the second half of the twentieth century showed that the experimental method is older than the Renaissance of the fifteenth or sixteenth century. Indeed, in one sense, the method is timeless. All animals, and consequently *homo*

sapiens from the inception of the species, have used an experimental method; that is to say, hypotheses are formed and tested. A young child warned that a saucepan is hot, lightly touches it to test out the hypothesis, then adds a piece of knowledge to strengthen the general law that things recently removed from the top of a hot stove maintain their heat for some time. Certainly all civilisations have made experiments and not least among them the Greeks.

It would be more accurate to rephrase Einstein. We are talking about degree, rather than an absolute change. We are all experimentalists, but some are more experimental than others. While experiments abound in everyday life, it is undoubtedly true that in many civilisations, and often increasingly over time, it was thought that knowledge of nature was already sufficient. All that we needed to know was known: the Buddha, Confucius, Muhammad and Aristotle had already provided the answers. So why experiment? Indeed, many of those in power would contend that one should not do so, for either it was a form of blasphemy or suggested doubt of the accepted wisdom on which rulers based their right to rule. This reminds us of the great obstacles which have almost always been placed in the way of a continuous development of human understanding of the natural world.

As Karl Popper pointed out some time ago, the Open Society, that is a world of continuous investigation and assessment of nature and social relations, has many enemies. Most human beings prefer certainty and order above all else. Most innovations and change threaten such orderliness. In particular, new ideas can be subversive and dangerous. Much of history shows the tendency of thought systems to close down, solidify and put up increasing barriers to any disturbance.

One aspect of this is what might be termed roughly the tendency towards inquisitorial thought. After a period of innovation and excitement, whether in fifth-century Greece, eleventh-century Islam, twelfth-century China or fifteenth-century Italy, when conceptual schemes are challenged and 'openness' reigns, there is usually a reaction. This is instituted by those who have never lost control of the thought systems – the Catholic Inquisition, the Communist Party or whatever closed system was in place before. 'Heresies' are now rooted out; challenges to the thought system are also seen as threats to the social and political order. The thought police are active, but do not need to be called in because of self-emasculation by individuals under all sorts of pressures, including those of loved ones.

The idea that, once 'open', systems of thought will increasingly become freer, is a naive belief which we can no longer accept. There are usually more people who have a vested interest in the intellectual (and technological) status quo, than those who have an interest in changing it. Such people also, characteristically, control the means to knowledge through controlling the educational system. Whether it is the Jesuits or the Mandarins or the Mullahs, a strict enforcement of the notion that certain ideas must not be challenged becomes widespread. It is more important to learn the old truths and reinterpret them than to learn new ones.

In fact, as the system closes down, it becomes almost impossible to challenge it in any way. Firstly, the new thoughts cannot be thought. If by accident they are, they will quickly be crushed. The central feature of the system, that is its systematic and absolute nature, compounds the difficulties. It is made up of a series of rigid rules and logical links. To challenge one part is therefore to challenge the whole – as in certain branches and phases of Islam, Roman Catholicism and Confucianism. What is odd are those few cases where the normal tendency towards rigidification does not occur and the system remains open and indeed becomes increasingly so.

A second difficulty, which compounds the political one, is what one might describe as the Brahmin or Mandarin trap. One of the major developments in mankind's history is in the improving technologies of the mind – the increasing power conferred by written scripts, new symbolic systems in mathematics, new philosophical systems. Already by the fifth century BC, notably with the Greeks, but also in China, India and the Middle East, very powerful tools of thought were available. Why then did it take so long for them to further deepen our understanding of the world about us?

One reason seems to be that there is a tendency for the literate group to become more and more obsessed with the form of the intellectual tools and passing these on, rather than their contexts and uses. The means, the tools of thought, became ends in themselves. Hence one sees the development of rote learning, endless chanting and repetition, the obsession with passing down the heritage unchallenged. In other words there is a routinisation and bureaucratisation of knowledge. Everything becomes codified and is deadened in the process. The aim is to affirm the received wisdom, rather than to question it. The disputations and confrontational logic of the Greeks combined with the social structure at their rediscovery in Europe avoided the full extremes of this consequence. But it was a very powerful tendency and one can see it well manifested in the later drying up of Arabic and Chinese thought. Minds were increasingly trained as memory stores, repositories of semi-ritual or religious truths. The idea was not to search for new truths, but to elaborate and fill out the old ones.

A further obstacle strengthened these general disincentives. The individual who experiments has little chance of success. The tools and techniques for discovering new things are weak, the complex interplay of causation too complex and concealed for the naked eye or the brain alone.

So, what is needed to make knowledge cumulative and to usher in a new, open world of reliable information? What can make an experimental method not merely a private matter of individual survival but a widespread and accepted method for reinterpreting the received wisdom about the world? There are probably many things, only a few of which can be lightly touched on here. Some of them become obvious if we look at the two bursts of experimentation which form the background to the era of Galileo.

The first wave is the work of Arabic scientists from the ninth to twelfth centuries. Three constituents of their world can be noted. There are the new theoretical frameworks from other civilisations which need to be absorbed by way of testing and exploration. The recovery and absorption of Greek and to a lesser extent Roman discoveries is but a part of this. They were also absorbing ideas from the east, most notably, the mathematics of India and the enormous learning of China. The great thinkers in this fertile area fed by so many streams of thought were faced with absorbing some new theories into their philosophical systems. The powerful Islamic civilisation that came to straddle much of Eurasia was placed at a perfect point between east and west, and created a dynamic whirlpool propitious for experiment. A sense of wonder, surprise, puzzlement, of new things emerging, all of which are essential for science, was there. Yet if this had been all, it is doubtful whether it would have led to much more than translations and annotations and some developments in mathematics and general theory.

Experiments need equipment, both mental and practical. After the collapse of the Roman Empire, the Arabic region was the glass-working centre of the world. By the ninth century, the most exquisite glass of all shapes, sizes and colours was being made. If an investigator wanted a glass flask to test out chemical theories, there was no problem in making it. If glass was needed to bend light in certain ways, or to break it up and examine its constituents, or to magnify the hitherto unseeable, or to experiment to see whether vision came through light that came into or went out of the eye, all was available. Glass provided, along with mathematical and logical tools from India and Greece, artefacts that made some experiments possible.

The Arabic thinkers had available reflecting devices, mirrors and refracting tools which were primarily glass bubbles with a liquid such as water in them. The Romans had used these water-filled glasses and both Pliny and Seneca had referred to the use of a sort of lens that has a naturally spherical surface, being a blown bubble, and hence not needing the polishing which creates a normal lens. The Arabs were heirs to this device. These water-filled glass globes can readily show magnifying effects, the concentration of radiation from the sun. So they can be used as burning glasses. Furthermore, if two glass globes, one of say four centimetres diameter and the other of ten, are filled with water and the smaller globe placed close to the eye and the larger one at a little distance from the smaller, then a (decidedly fuzzy) inverted, but nevertheless quite reasonable image of distant objects, somewhat magnified in size, can be seen. The two globes can very loosely be called lenses and are quite easy to prepare where there is a tradition of glass-blowing and the availability of clear glass, as was the case in several Islamic centres.

Such tantalising inverted images haunted the imagination of thinkers from the eighth century to the sixteenth, and gave rise to remarks such as that of Roger Bacon that 'the most remote objects may appear just at hand' and have led later commentators to think that medieval Arabic or western experimenters possessed some form of telescope. In a way they did have such a device, because the magnified image is real enough. Yet the image is so distorted as to be useless in preparing a practical device. That would have to await the lenses of the early seventeenth century.

Glass instruments were vital in the period when Islamic experimentation was at its peak, between the ninth and twelfth centuries. This is evident if we look at those fields where the Arabic thinkers made their greatest contributions. In medicine, the use of glass to see the minute or to test compounds is central. In chemistry, one of the greatest Arabic achievements, glass tubes, retorts and flasks are essential for the laboratory. In optics, which in turn deeply influenced physics and geometry, we know of the role of prisms and mirrors in Arabic work. They could also produce a colour spectrum through the dispersing effect of the naturally found quartz crystal. It also seems possible that they used plano-convex lenses and there are suggestions to this effect in the texts, though it cannot be substantiated by the evidence of any remaining artefacts.

The first major thinker on optics was Al-kindi (c. 801–66) who worked on a theory of light and put some order into the chaos of observations and the relics of Greek science. In about 984 Ibn Sahel wrote a treatise on burning-glass and other kinds of mirror. His proof showed that he had a mastery of geometrical reasoning, which was a substantial contribution to the later sciences of optics and physics in general. This is very important, firstly because it is correct, secondly because it had been achieved by the application of geometrical reasoning to a physical process, thus laying the groundwork for the use of mathematics in optics. Of course, it requires a further assumption, namely that light travels in straight lines, an assumption which they had and is indicated by daily observation of light beams as they travel through clouds or holes in a shutter. All of this set the tone for rigorous logical thinking on the back of experimentation, which is really what the generation of increasing reliable knowledge, also known as science, is all about.

Possibly the greatest of the Arabic philosophers of light, Alhazen, born about 965, worked in Cairo making copies of Euclid and Ptolemy. He died in about 1041 having published some 120 books. His work on optics was translated in about 1200 into Latin. *On Vision* was printed in 1572 and dominated speculation until Kepler's revision in 1610. It was an empirical work, drawing conclusions from what Alhazen had observed. For instance, he did a famous experiment where he placed three candles on one side of a screen with a small hole in it, and observed the points of light which they threw on the wall on the other side of the screen, thus showing that light rays travelled in straight lines and that as the rays passed through the small hole they did not get absorbed into each other or bent in any way. He argued that the incoming form was purely visual; recognition is the result of memory and inference. Shapes and colour come into the eyes, rather than, as in the old theory, light travelling out of the eye to find objects. He suggested that the surface of an object consisted of many different points or specks which we see and then rearrange. He analysed an old subject in a new way and contributed to an understanding of the function of the eye. However, he did not perform a dissection of the eye – Islamic law forbade it – and hence had an erroneous picture of its anatomy. He may have invented the *camera obscura*, and certainly used one.

Yet despite the enormous advances made by Arabic theoreticians, it is generally admitted that they did not break through into that set of interconnected practices which we call science. Those who have looked carefully at the achievements of Arabic scholars are agreed that, for some reason or another, they fell short of the breakthrough that occurred in western Europe sometime after the thirteenth century.

The Arabic scholars clearly knew about plano-convex pieces of glass and used them, though they did not apparently use double-sided lenses. They discussed spherical and parabolic mirrors, the *camera obscura*, plano-convex lenses and vision. Likewise the medieval European writers clearly knew about prisms and plano-convex lenses and used them in their experiments. While it is the case that there is little evidence that the lens with both sides curved was developed before 1280, it seems over-restrictive to argue that the lens was unknown before this date, and it certainly gives a misleading impression when we consider the accounts of the effects of glass on magnification by medieval scientists, which were to have such a profound influence. If, as we shall see, one of the most revolutionary effects of glass was in telescopes and microscopes, the seeds of all this are to be found in the thirteenth century. It is worth quoting the famous accounts by two thinkers on the potentials of glass as a way of seeing new things. Both of them show very well the way in which early thinkers were tantalised by the awareness of some extraordinary properties in glass which they could not fully utilise because of the state of the technology.

Robert Grosseteste (c. 1175–1253) in his *Perspectiva*, on the effects of glass

> shows us how we may make things a very long distance off appear as if placed very close, and larger near things appear very small, and how we may make small things placed at a distance appear any size we want, so that it may be possible for us to read the smallest letters at incredible distances, or to count sand, or grains, or seeds, or any sort of minute objects ... It is obvious from geometrical reasons, given a transparent body (*diaphanum*) of known size and shape at a known distance from the eye ... all visible objects may be made to appear to them in any position and of any size they like; and they can make very large objects appear very small, and contrariwise very small and remote objects as if they were large and easily discernible by sight.

Roger Bacon (1214–94) took these ideas further, for he now had available the work of Alhazen. He wrote that

> If the letters of a book or any minute objects be viewed through a lesser segment of a sphere of glass or crystal, whose plane base is laid upon them, they will appear far better and larger ... And therefore this instrument is useful to old men and to those that have weak eyes. For they may see the smallest letters sufficiently magnified ... the great-

> est things may appear exceedingly small, and on the contrary; also that the most remote objects may appear just at hand, and on the contrary. For we can give such figures to transparent bodies, and dispose them in such order with respect to the eye and the objects, that the rays shall be refracted and bent towards any place we please; so that we may see the object near at hand or at a distance, under any angle we please. And thus from an incredible distance we may read the smallest letters, and may number the smallest particles of dust and sand, by reason of the greatness of the angle under which we may see them; and on the contrary we may not be able to see the greatest bodies just by us, by reason of the smallness of the angles under which they may appear. For distance does not affect this kind of vision, excepting by accident, but the quantity of the angle.

Although the application of this theory had to await the transformation of spectacle lenses into telescopes and microscopes, the idea was now established that glass could open up new realms of knowledge, the microscopic and macroscopic. The immense potential of lenses, partly developed by Arabic thinkers, was beginning to be realised.

The speed of the inrush of new knowledge into medieval Europe is well known. Almost all of the great tradition of Graeco-Roman science had been lost or garbled after the collapse of the Roman Empire. Although little pieces were retained, perhaps up to three-quarters or more had been lost. In the twelfth and thirteenth centuries the great burst of translations, many of them from the Arabic, transformed knowledge in western Europe. Coincident with the founding of universities and the flourishing of the Church and the economy, which in their different ways provided the institutional infrastructure for the new learning, new knowledge flooded in. For not only were the astonishing achievements of the Greeks and to a lesser extent the Romans, particularly in the latter case in natural history, engineering and medicine, made available, but what reached Europe was now given added force by the achievements in synthesis and extension attained by the Arabic scholars. They had absorbed much of the accumulated knowledge of China and India as well, in particular in relation to a better mathematical system, and then added their own experimental and theoretical observations.

Within a period of about one hundred and fifty years western Europeans moved from a world where even the best informed knew little about the principles of the natural world, just what had been preserved in a few monasteries plus some native ingenuity, to one where they had before them much of the accumulated reliable knowledge that had built up over most of Eurasia during three thousand years. The excitement, the stimulus to questioning, the wonder and curiosity are palpable in the great thinkers of the time, and perhaps nowhere more so than in the works of Roger Bacon.

This curiosity, the impetus to test and speculate, the sense that there were expanding horizons of knowledge, that not all was known and there were new worlds to be discovered, were boosted by the rapidly expanding wealth and technology of the period. The new burst of power through the intensive exploitation of wind, water and animals, the growth of trade and cities, and the expansion of Christianity, which was temporarily at least sympathetic to the study of God's law, all encouraged experimentation and the harnessing of new knowledge. The symbol and expression of this expansion lies in religious buildings, in the magnificent development of the Gothic cathedrals. These very cathedrals also show us once again the necessary counterpart which allowed wonder and curiosity to be turned into progressive experiment.

As we have seen, from about the late twelfth century a never-forgotten tradition of glass-making in Europe blossomed. This development took place most famously in northern Italy, but also in most other parts of western Europe. The art of the glass-maker, in turn fed by the new knowledge in geometry and optics, flourished and was increasingly applied to uses of glass explicitly designed to improve the human eye and what it saw.

One example of how complex the interplay between the new glass tools and abstract knowledge can be, is seen in the development of medieval mathematics. At first sight this seems quite distant from glass. After all Arabic mathematics, in particular arithmetic and algebra, which had such an important influence, came from a more or less glassless civilisation, India. Yet it is significant that Einstein singled out not mathematics but (Euclidian) geometry as the key tool in the 'scientific revolution'. Perhaps geometry, in itself, is no more important than algebra or arithmetic. Yet without the progress of geometry, many of the great advancements in astronomy from Copernicus onwards would be inconceivable. It is well known that geometry was not greatly developed in China. Likewise, while the Greeks had laid the foundations for geometry, the subject came alive again and was enriched first by the Islamic and then by the medieval European mathematicians. This was not just a matter of recovering the lost Greek inheritance, difficult though this was. There was a conspicuous improvement in the understanding of space and light which lies at the heart of geometry.

These improvements were catalysed by developments in optics and specifically the extensive work on reflecting, bending and analysing light by Adelard, Grosseteste, Bacon and others. For this they used glass tools, particularly mirrors, prisms and lenses. Yet in order to sustain interest, to build up a community of interacting scholars, to have a sense of control and insight into hitherto intractable problems, the glass tools used in geometrical experiments were very important. Their role has disappeared, for once the discoveries were made such tools seem unimportant. It may all seem easy, perhaps inevitable, after the event. But setting out to test and improve Greek geometry it was essential for the great medieval philosophers and mathematicians to have at hand the new tools not available to the Greeks, if only to give them strength in the task.

In recent years there has been a growing realisation of the sophistication and importance of medieval optics. This partly stems from the work of A. C. Crombie. He showed the way in which research into the causes of the rainbow, using sunlight passing through a spherical glass, often a urine flask full of water which refracted and then internally reflected light, glass prisms, hexagonal crystals and so on, begun by Grosseteste, carried on by Albertus Magnus, Roger Bacon and Witelo over the thirteenth century was completed in the early fourteenth century by Theoderic of Freiburg. The whole of this investigation using glass was crucial in the development of two of the most important methodological underpinnings of modern science, the experimental method and the principle of economy (that nature works by the shortest and simplest route – famously known as Ockham's razor).

Particularly important was the work of Roger Bacon. Two of Bacon's works dealt with optics: *De multiplicatione speciarum* proposed a philosophy of natural causation based on an optical model and *De speculis comburentibus* investigated ways in which light was propagated and applied this to the analysis of the burning-mirror. These were among his most successful and influential works, centring on problems in geometrical optics and building, as we have seen, on the methodology introduced by the Arabs in their great contribution to the development of the sciences. All of this work depended on optical tools, most of them made out of glass. He looked at various curved surfaces and the principles of refraction and reflection from these surfaces, using concave and convex mirrors. He looked at clear mirror images to see how the image is reflected in the mirror. Mirrors, prisms and lenses allowed the new mathematics and geometry to develop.

There are a number of reasons why the expansion of reliable knowledge in medieval Europe took earlier ideas on to a further stage. There was more knowledge available to the early European scientists than to the Arabic thinkers, for on top of the revived Greek knowledge there was the added component of the completed Arabic synthesis. The speed of the inrush of new thought was far greater in western Europe. Spread over half a millennium in the Arabic world, it occurred in a third of that time in Europe. So the propulsion towards wonder and curiosity was greater. The shock of vast realms of new knowledge surfacing and flooding in must have been immense.

Likewise the quality of the glass available for experimentation was noticeably higher. Mirrors were increasingly made of glass, which gives a more detailed reflection of depth and colour than metal mirrors. The lenses which began to be used were able to provide hints of a world below the level of normal vision, the prisms were more sophisticated, the chemical apparatus improved as glass technology rapidly developed.

In fact, the laboratory without glass is almost impossible to conceive of; what would be in it (apart from books and a few measuring tools), without the retorts, flasks, containers, mirrors, lenses, prisms and so on? When we look at medieval descriptions of the working places of scientists in the west, they are often filled with glass devices. We are beginning to appreciate how widespread these glass instruments were. For instance, recovered medieval English glass includes a wide range of chemical equipment. The laboratory equipped with glass did not develop outside western Europe and, to a certain extent, the Islamic world.

As we have seen, one of the rapid developments in glass technology was the making of panes of window glass, plain and coloured, which was particularly noticeable in the northern half of Europe. One very practical effect of this was on working conditions. In the cold and dark northern half of Europe people could now work for longer hours and with more precision because they were shielded from the elements. The light poured in, yet the cold was kept out. Prior to glass only thin slivers of horn or parchment were used and the window spaces were of necessity much smaller and the light admitted, dimmer.

It could also be argued that windows altered thought at a deeper level. The question here is the way in which glass, whether in a mirror, window, or through a lens, tends to

3. Glass, alchemy and chemistry

Glass has been an essential material for testing the properties of various substances throughout history since it is inert and does not affect the experiment. At first it was particularly important in alchemy, as *The Alchemist* by Johannes Stradanus (in the Palazzo Vecchio, Florence) shows in a workshop filled with glass instruments. Later it was an essential material for apparatus in the chemical experiments which formed a central feature of what we call the scientific revolution.

concentrate and frame thought by bounding vision, and at the same time leads to abstraction and attention to the details of nature. It seems likely that the glass window altered the relations between humans and their world in ways which it is now difficult to recover. It may have encouraged the contemplation of external nature from within the house, an appreciation of nature for its own sake, seen through a window. Yet clear glass was only one part of the way in which windows became magic casements. It is rather noticeable that all the greatest medieval scientists in the west were churchmen: Adelard of Bath, Pecham, Grosseteste and Bacon. Although this may be due to the fact that only ordained clerics had access to the time and learning to make a high level contribution, it still remains interesting that they should have turned their attention so strongly to optics and related subjects. Is it just a coincidence that they were living at a time when the new cathedrals were being built? It seems probable that the light that flooded in through the magnificent stained glass windows influenced them.

No wonder optics became a central field of medieval science in the west, the counterpart of physics in later centuries. The metaphysics of light, its symbolic importance both in Greek Neo-Platonic thought and in Christian thought, is a rich theme with enormous consequences. It is also immensely complex. Styles of thought were inherited, but given a new impetus by the expanding world of light through the windows in churches and private houses. Thus light and knowledge, truth and beauty, became fused. It was glass that united them. So there was a combination of the impetus to explore and a developing set of glass artefacts. This made that exploration possible and became part of what we call the experimental method.

4. Apparatus used by Priestley

Joseph Priestley was one of the leading natural scientists of the eighteenth century. This represents the typical apparatus used in the early study of both the physical and chemical properties of air and other gases. Such experiments would not have been possible without the use of glass tools.

Of course, in this discussion it is important not to fall into the trap of believing that glass always led to a closer approximation of what finally turned out to be correct knowledge as we conceive of it. There were many fruitful errors on the way. One of the most important roles of glass was in 'natural magic', especially in alchemy and astrology. Alongside the curiosity and desire to understand God's laws of a man such as Roger Bacon, there were numerous people who desired power through the making of wealth (alchemy – the search for gold) or foreknowledge of the future (astrology and sooth-saying). For them glass was a powerful tool and retorts, mirrors and lenses were developed in this fermenting no-man's land, one of whose last great magi was Newton. For example, the extensive use of mirrors in magic, from the early fourteenth to the end of the sixteenth century, is widely known from the work of Giordano Bruno and the Hermetical Tradition, the Rosicrucians, Della Porta and John Dee. Research in this field shows that the old opposition between science and magic is being rethought.

Yet if we try the thought experiment of wishing away glass in Islamic and medieval Christian civilisation, it is not difficult to see how reliable knowledge would probably have come to a halt. Any child will tell you that an exciting science book is not enough, even when it sets out all sorts of possible connections and theories. Only when we are equipped with a jam jar, a magnifying glass or test tubes and microscopes, will the amazing world of nature's secrets be unlocked. Obviously glass, on its own, is not enough. Without the burst of curiosity and new knowledge from ancient and Asian civilisations, all the glass in the world would probably have had little influence on thought. It is the combination of curiosity and tools that is important. Of course, there were many other factors which have often been pointed to. The growing explorations along the land routes to Asia, the demands of competition and war, the growth of cosmopolitan cities, the rise in wealth, the development of universities in the west and so on. Yet glass, it seems to us, is a *sine qua non* of the development of the experimental method we call science.

Science consists of verifiability, repeatability and openness to refutation. The pure speculations of many thought systems were not open to such checks. Plato or Confucius or the Buddha set up systems which were internally consistent, coherent and closed. They could not be challenged from within nor destroyed by 'evidence'. Nor could the casual observer check their parts. The logical experiments could not be done again. It would be as meaningful to 'test' them with experiment as it would be to 'test' the Mona Lisa, Chartres cathedral, Handel's *Messiah* or Shakespeare's *Hamlet*. They were statements which could not be verified. Modern science, however, depends on the formulation of laws based on experiments which can be repeated by others.

Glass shifts authority from the word, from the ear and the mind and writing, to external visual evidence. The authority of elders is challenged; the test is the individual eye and the authority of the doubt-filled and sceptical individual. The primacy of demonstration through showing something happening became obvious. One must test every piece of received wisdom by the evidence of one's eyes. What others see in an experiment, which is potentially reproducible, is more important than what is asserted by authority (the word).

Thus it could be argued that glass helped change the balance of power from the mind to the eye. The frequently noted empiricism and positivism of the west, where only seeing is believing and demonstration is essential, became distinguishing features of the new cosmology. Every time the technology of seeing was improved, it lent more authority to the experimental method. It confirmed the view that God had created a mysterious, little understood world, yet one which contained clues to certain general rules or principles which could be known and once established could be used to base other findings on. There was no fixed and known pattern in the mind, just divinely inspired curiosity for which the new tools, including glass and mathematics, provided the data. No longer reliant on thought experiments, one looked at nature, from every angle, at the minute and macro levels, sideways and upside down, with mirrors, lenses and prisms, under different conditions of heat and cold and in various mixes in glass vessels, to torture it to see what it was made of.

The shift from the authority of texts, and of received wisdom, to the authority of the eye and the perception of each observer is one of the most intriguing aspects of what happened. It is possible to wonder about the role of glass in giving authority to the experimenter or author and his vision. Many have talked of the final rejection of Aristotelian philosophy. It is tempting to argue that the products of a new knowledge substantially based on glass finally overthrew it. In order for modern science to emerge the overthrow of Greek science was a condition. It does not seem implausible to argue that this massive task could not have been achieved without the confidence which glass produced. The evidence lies in the war between the Aristotelians and what they considered to be the lies, deceits and false knowledge created by glass.

It can also be studied in the great shift in self-confidence and endorsement of the authority of the individual and of vision, which many term the Renaissance. For the next great expansion of reliable knowledge in the west did not occur in the universities and in philosophy or what we would call science, but in the arts and engineering, and in particular in architecture, painting and drawing. We can look at the influence of glass in relation to the two major features of the Renaissance. One is the understanding and representation of the natural world; the other is the changing concept of the individual.

4

Glass and the Renaissance

The eye is the commander of astronomy; it makes cosmography; it guides and rectifies all the human arts; it conducts man to the various regions of this world; it is the prince of mathematics; its sciences are most certain; it has measured the height and size of the stars; it has disclosed the elements and their distributions; it has made predictions of future events by means of the course of the stars; it has generated architecture, perspective and divine painting. Oh excellent above all things created by God!

Leonardo da Vinci, *On Painting*, 21

HUMANS ARE NOT CAMERAS. They do not naturally see the world as it is, but as they expect it to be. They see with blinkers on, as it were, selecting bits and pieces they can comprehend. This view, which has been endorsed by more recent work on the psychology of perception, suggests that we do not see the world straight on but unconsciously distort it, or rather reinterpret it. A flow of light comes into the eyes, but what we see has to be created from the meaningless jumble of colours and shapes. For example, the image always strikes our retina upside down and we have to turn it over, which we have learnt to do. We use another trick to make sense of mirrors. Now these perceptual constraints are so powerful, yet beyond our control, that they actually make the world into the shape which we expect to see. We are like Wittgenstein's famous fly trapped in the constraining fly-bottle. We are systematically biased. We immediately start to interpret the world, even before the light enters our eye.

This set of constraints is reinforced by another when we come to represent what we see to others. The result can be seen if we look at any of the great art traditions of the world up to about AD 1250. If we survey South American (Aztec/Inca) art it is conventionalised, symbolic, two-dimensional, without perspective. It is more akin to a cartoon or to a form of writing than to modern western realist painting. The same is true of the other non-Eurasian art traditions in Australasia and sub-Saharan Africa. If we are tempted to think that this has something to do with their pre-civilisational (i.e. pre-writing) situation, we will soon be disabused by looking at the civilisations of Eurasia itself. The early civilisations of Mesopotamia and Egypt, after the invention of writing, have a similar flat, stereotyped art. The art of Greece reached wonderful levels, particularly in sculpture, but the painting shows little sign of the correspondence to nature of modern western art. The majority of Roman art, though having representations of some depth, still lacks true perspective of the kind that was developed in Europe after the fourteenth century.

The development of the various non-western art traditions is particularly revealing since they did not collapse, unlike Greece and Rome or the early empires of the Middle East. So we can watch their development over the crucial period from the time of the fall of Rome in about the fifth century up to the eighteenth century. In Byzantium, the iconic and non-realist art using conventional symbols continued largely unchanged until the fall of Constantinople in 1453. In Russia, the same was true until the eighteenth century. In Arabic societies, soon dominated by Islam, there is no sign of a fundamental change in artistic representations. We can see superb illustrations of the world in the Mughal art of the court of Akbar and Shah Jehan in India in the seventeenth century. Wonderfully detailed, the paintings appear largely flat, shadowless, without accurate perspective or pictorial space.

The art of China and Japan is equally intriguing. It is exquisite, often very finely crafted; yet its essence is, when compared to post-Renaissance west European art, on the same side as all the traditions we have looked at. It almost always lacks perspective depth, it lacks realism, and the backgrounds are often impressionistic. It appears to be very highly stylised. To an outsider, it seems to be painting by a code, employing images as symbols which the viewer can interpret as references to something else. Paintings largely seem to be mnemonic devices, reminding the viewer of emotions, but not systematically exploring the world of nature. This tradition, both different from and overlapping with that in the other areas we have looked at, is one which continues with only minor modifications until at least the eighteenth century and later.

Thus we see that the way in which civilisations up to about 1250 visually represented or explored their world was of a particular nature. They all tended to use paintings as anagrams for the world. It is almost as if the painting was a written language in which the power is derived from the arbitrariness of the symbols. The gap between the signified (the world of nature) and the signifier (the artistic representations) is often very great. The moon, a twig, a leaf stand for an immensity of meanings in a Chinese or Japanese painting, as they do in a poem. The paintings are like visual poems, composed according to anciently established rules, turning the artist inwards on himself and away from the outside world. They are concerned with symbolic analogies, with a cosmology where inner and outer essence and form are closely connected.

It is probable that much of this was deliberate, that, as with alphabetic writing, artists had discovered that the more conventionalised the symbol, and the better educated their audience, the more powerfully they could stir their feelings. Yet as well as this, there seem to have been other pressures, and we can see this clearly if we reconsider what our basic problem is.

If we examine the history of representations in early civilisations we find that the story is not as simple as suggested above. Firstly, there is evidence that Graeco-Roman art had developed quite good perspective in various representations, and then this was lost for a thousand years. Secondly, there is the evidence of quite realist representations in some of the early paintings in the Ajanta caves of the fifth century in India. Thirdly, there is the case of China. For example, the famous eleventh century set of drawings, including boats 'going up the river', which shows a grasp of some of the rules of perspective. This technique was largely discontinued over the next few hundred years. Fourthly, there is the case of Giotto in early fourteenth-century Italy. His ideas were little developed for nearly a hundred years, to be rediscovered and elaborated in the fifteenth century.

5. Going up the river

Detail of a handscroll. Ink and slight colour on silk. Palace Museum, Peking. This famous scroll of life along the river shows that the Northern Sung painters had mastered realistic drawing, shading and foreshortening by the twelfth century, only to abandon those techniques later as too realistic and not suitable for a scholar-painter.

If we consider these examples, we are led to reflect more deeply on the nature of human sight. We know that, with their binocular vision, humans see the world in perspective. We know that things appear to get smaller as they are further away, vanish towards a horizon and so on. A child intuitively knows all these things and *homo sapiens* would not have survived for long if they had been forgotten. We are also able to paint or draw the world in quite reasonable perspective if left to our own devices, as children sometimes do, and the young shepherd boy Giotto famously did.

So we must turn the question on its head. Perspective and realist portrayals are natural and normal, but usually the cultural conventions of a society teach artists and others that the representations their audience want are not of this kind. Artists are, as it were, systematically taught to distort the world they see and would normally portray, in order to make it fit a symbolic system which conveys deeper meaning than the prosaic world of sight. What is the point of art if it merely duplicates what the eye sees? Putting the problem this way leads us to speculate on what cultural pressures have been erected to stop most art traditions from striving for realistic, perspective-filled representations. We would then be led to ask what it was that so forcefully dislodged these pressures so that for a brief period in one civilisation (western Europe from the fifteenth to the nineteenth century) a realistic form of art came to dominate. What turns rather isolated cases of perspective and realist representations such as those of the early Chinese artist of the eleventh century or Giotto into a vast movement which then transforms the whole of human vision and reality?

What seems certain is that it would require a considerable jolt to push one civilisation away from what appears to be a common sense and obvious vision of the world, and hence the only and necessary way of representing it. The fact that no civilisation had been able to break out of the fly bottle before 1250, and that, to a considerable extent, the most sophisticated and artistic civilisations of the world – Islam, China, Japan – never did so from within, is indicative of the strength of the tradition. What force could then be strong enough to shatter the glass? Or at least, what could make the fly free by making it aware of the invisible constraints on its vision, for which it could then create some artificial compensation?

Obviously, the first thing to do is to show that the revolutionary change towards accurate perception and representation of the natural world did, indeed, once happen, and where and when it happened. The case is well known and is one of the most famous episodes in history. If one looked across western European art in the eleventh and twelfth centuries, it was essentially similar to all those other traditions we have described. It was iconic, mainly religious, with strong symbolism, flat, stereotyped painting by a code. Although the content was different from Islamic or Chinese art, the aim, to remind people of other things, to present a set of linked symbols, to explore inner emotions rather than the material world, was similar. It was as far away from naturalism, from a photographic representation of the world, as was any other art tradition. Nor were there any obvious signs of something different that was about to emerge. Then, in the two hundred or so years between 1300 and 1500 there occurred a revolution in seeing and representation to which we ascribe the label 'Renaissance', which set one part of the world off on another trajectory.

The revolution, particularly in painting and architecture,

with special emphasis on pictorial space and perspective, is the subject of libraries of books. Many believed that Giotto effected the first transformation, the ability to move from what Gombrich calls 'picture writing' to painting with some depth, for he had 'rediscovered the art of creating the illusion of depth on a flat surface'. Although revolutionary, this was an intermediary stage between the earlier Graeco-Roman art and a new realism which suddenly emerges in the period from roughly 1400 to 1500. This revolution brings in the rules of perspective, and with them a host of new technical methods. If we look at the art and architecture produced after about 1400 simultaneously in south and north-western Europe, from Alberti, Brunelleschi, Masaccio and later Leonardo in the south, and from Van Eyck and Rogier van der Weyden to Pieter Brueghel in the north, it is an enormous shock. It is as if the world has suddenly been uncovered, a layer stripped away. There is clarity, interest in detail, a mirror-like accuracy of man within his world. The quantity of reliable information purveyed by a picture is increased enormously. Pictures no longer serve principally to remind or signify, they open magic casements upon new worlds. It is like looking through a strong lens at the world. The world is often richer and brighter in the picture than it is in reality. It is as if it is seen through a magnifying glass.

So why and how did this revolutionary change occur in only one civilisation – from Italy to the Netherlands – where artists saw and represented man's physical surroundings accurately for the first time? Here we are in for another surprise, since it appears that we lack a plausible explanation for one of the greatest changes in human history. Many commentators on Renaissance art and culture from the later nineteenth century onwards have documented the revolution, described its quintessence in changes in perspective and concepts of space and precise delineation of nature, but rejected the revival of classical learning as its main inspiration. So they leave us with another description of what happened but not why. The origins of the Renaissance are obviously an immensely complex problem, and there are likely to be numerous chains of causation. Our task here is just to look at one possible cause, which has often been ignored, that is the role of glass in providing both the jolt and technical support for a realignment of vision.

While it is true that we are aware of perspective, and children without lessons can represent the world quite realistically in perspective, there is more to it than that. It is difficult to turn such knowledge and fairly elementary ability into a convincing representation that will give others the illusion of real space and form. To proceed far requires consciously making the implicit more explicit. But while it is difficult, it is not impossible for a person to work out the method for themselves, as our exemplars including Giotto have shown. Yet to transform this achievement into a movement which will alter the world, as the Renaissance did, requires some added features.

One is an audience that wants to be 'deceived' into thinking that what is in fact two-dimensional appears to be three-dimensional. Here is one of the great differences. It is well known that Plato felt that realist, illusionary art should be banned as a deceit, and most civilisations have followed Plato, if for other reasons. For the Chinese (and Japanese) the purpose of art was not to imitate or portray external nature, but to suggest emotions. Thus they actively discouraged too much realism, which merely repeated without any added value what could anyway be seen. A Van Eyck or a Leonardo would have been scorned as a vulgar imitator.

In parts of Islamic tradition, realistic artistic representations of living things above the level of flowers and trees are banned as blasphemous imitations of the creator's distinctive work. Humans should not create graven images, or any images at all, for thereby they took to themselves the power of God. Again, Van Eyck or Leonardo would have been an abhorrence. Even mirrors can be an abomination, for they create duplicates of living things. So in considering the development of the extraordinary movement in the west, we always need to bear in mind the wider cultural setting. That is to say we have to consider the audience and the general view on the role and limits of artistic work.

The second factor is how the prospective patrons and customers themselves saw the world in their normal life. It is one thing to see an actual three-dimensional external world realistically; it is another to 'read' a two-dimensional, artificial representation of that world in a way which tricks the mind into suspending disbelief sufficiently to feel one is seeing a slice of reality. It is constantly stressed by art historians that the audience has to be taught to read realist (or any other) form of art. The audience also has to want to be tricked by artists. So in order for such art to spread, two things have to occur more or less simultaneously. The artist has to educate his audience, to shape them so that they can read his work. At the same time, there has to be enough in what he portrays which is in conformity with their vision for the audience to feel attracted to his work in the first place.

Here we may be getting closer to an area where the

development of glass technologies began to have an effect. It is conceivable that the widening experiences of seeing the natural world more realistically and intensely, in better mirrors, through window frames, through spectacles in older age, may have tipped the balance just enough to change the view of the world. That crucial bourgeois patron group who bought and exhibited the new art now found it meaningful and attractive. Their experience made them feel more like cameras, and they adored the painters who captured for them the world on a glass plate. The square canvas on the easel and later on the wall was a movable window, opening upon an imaginary world, a magic casement taking the viewer out into any type of imagined space. It was the antecedent to the television screen.

The third factor is related to the development of the technology of enchantment, or illusionist technology as others name it. There are several aspects to this. One concerns the replicability of realist art. That is to say, how easy is it for someone who is not a genius like Giotto to be able to deceive other people's eyes into believing they are seeing the three-dimensional world on a two-dimensional surface? Another aspect is the need for an explicit methodology to make it possible to explore the more difficult areas of realism and to lay down the rules so that gains in knowledge are not lost. There is a long way between even the eleventh-century Chinese drawing or Indian painting and the work of Van Eyck or Leonardo.

Let us look first at the question of replicability or multiplication. As Alberti, Leonardo and Dürer, among others, fully realised, the new realist art would flourish only if its principles were made explicit in a set of manuals which would spread the techniques from geniuses to the merely talented. Not everyone was a Giotto or a Van Eyck, so it was essential that people understood, in detail, how to look at nature, how to translate what they saw on to paper or canvas, and how to deceive the eye of the beholder into seeing the same thing. The rules they had to learn, as we see from the manuals, were largely mathematical, concerned with the properties of light and the nature of the eye. An artist had to take a course in basic geometry and optics. And where had these rules come from? The writers of manuals openly acknowledged that they had come from Greek sources, through Arabic scholars, and been made useful by medieval scientists like Pecham, Grosseteste, Bacon and Witelo. To what degree these later authors realised that glass in the form of lenses and mirrors had been an essential tool for these philosophers is not clear. Yet there can be no doubt that at this point there could not have been a set of rules for them to follow without the philosophies developed with the aid of glass.

A further aspect of the technology of enchantment was the elaboration of tools to make the exploration of reality more complete. This is particularly obvious in the work of Alberti and Leonardo, where the same foundation of earlier geometry and optical knowledge is used to work out stratagems to make paintings and buildings in a new way. Although Giotto and some Roman, Indian and Chinese artists had achieved a great deal with craft skills, the extraordinary realism and accuracy of the greatest of the Renaissance artists required further serious thought about the properties of light and space. This then demanded advanced geometry and knowledge of how the eye might work. This in turn depended on the flourishing of a glass-enhanced science of geometry and optics in medieval Europe. Without this, it is very difficult to see how the Renaissance could have occurred.

The fourth area lies in the tools that an artist could use. The development of oil paints associated with Van Eyck allowed depth and richness of texture and colour, but we would suggest that equally essential, though less noticed, was the development of tools of glass. These glass tools were important in two major ways. The first was to provide a shock, corrective or extension to the eye. The main tool here was the mirror. As many have pointed out, we tend to become too familiar with the world around us. The mirror, by reversing it, throws it into a new light and, in a curious way, makes it more intense. We, and particularly artists, see the world differently with a mirror.

Filarete, a fifteenth-century contemporary, wrote of the discovery of the laws of perspective by Brunelleschi: 'If you should desire to represent something in another, easier way, take a mirror and hold it up in front of the thing you want to do. Look into it, and you will see the outlines of the thing more easily, and whatever is closer or further will appear foreshortened to you. Truly, I think this is the way Pippo de Ser Brunellesco discovered this perspective, which was not used in other times.' That the mirror was the mother of Renaissance perspective is a theme taken up by Samuel Edgerton in his *The Renaissance Rediscovery of Linear Perspective*. He carefully traces the various converging sets of ideas from Greek and Arabic philosophy through medieval optics, geometry and cartography which led to the fateful moment in the Piazza del Duomo in Florence in 1425 when Brunelleschi made his major discovery of the laws of perspective. Mirrors had been standard in artists' studios for several hundred years, for example Giotto had painted 'with the aid of mirrors'. Yet Brunelleschi's extraordinary breakthrough is the culminating moment. Without what Edgerton calculates to be a twelve-inch-square flat mirror, the most important single change in the representation of nature by artistic means in the last thousand years could not, Edgerton argues, have occurred.

Leonardo called the mirror the 'master of painters'. He wrote that 'Painters oftentimes despair of their power to imitate nature, on perceiving how their pictures are lacking in the power of relief and vividness which objects possess when seen in a mirror ...' It is no accident that a mirror is the central device in two of the greatest of paintings – Van Eyck's 'Marriage of Arnolfini', and Velazquez's 'Las Meninas'. It was a tool that could be used to distort and hence make the world a subject of speculation. It was also a tool for improving the artist's work, as Leonardo recommended. 'When you wish to see whether your whole picture accords with what you have portrayed from nature take a mirror and reflect the actual object in it. Compare what is reflected with your painting and carefully consider whether both likenesses of the subject correspond, particularly in regard to the mirror.' He elaborates on this as follows: 'You should take the mirror as your master, that is a flat mirror, because on its surface things in many ways bear a resemblance to a painting. That is to say, you see a picture which is painted on a flat surface showing things as if in relief: the mirror on a flat surface does the same.' The aim is to paint so that the picture looks 'like something from nature seen in a large mirror'. Finally, it gave the artist a third eye, an eye on a stalk as it were, so that he could see himself. Without a mirror, the great autobiographical portraits, culminating in the series by Rembrandt, could not have been painted.

6. Dürer's drawing device

The outline of the subject and the surroundings is traced with a crayon or with paint on to a clear sheet of glass. This outline is then used for filling in the picture. The eye has to view both the glass and the subject from a fixed position and an adjustable device, using a small hole, which enables this to be done is shown.

It may be that mirrors increase the intensity of human sight in other ways as well. Seeing is a dynamic process. If we stare directly at an object for a long time,.we cease to see it. Only if we change our angle of vision, sweeping the eye across the object, do we continue to see it. Mirrors help us to see clearly for, whether held in the hand or altered by the movement of the person who gazes into them, they increase the amount of movement which is projected on to the eye. Furthermore, they are often inside dark rooms, reflecting the brightness of the outside world. The eye that has compensated for the dark surroundings sees the world all the more intensely in the mirror, just as a television set is much more effective in a darkened room.

A second way in which glass is important is as a tool in framing and fixing reality. This is the area where panes of glass rather than mirrors are important. A painting is a window opening on to another reality, and the window here is not just a metaphor. The magic casement through which one looks into three-dimensional space was an increasingly everyday experience of Europeans from the fourteenth century as good window glass spread. In a civilisation where oiled or mulberry paper was used, as in China or Japan, the idea of sitting inside and looking through a small, painting-sized aperture which focused a scene would hardly be developed. Either the wall was taken away completely, as with a *shoji* or window screen, and one was, in effect, outside, or one was inside with whiteness between one and the outside. Yet for richer Europeans, houses became like camera lenses or peep shows; one sat in muted light and looked out on the richness of colour. Or, as in Dutch interiors, one looked into rooms filled with light through the windows. Whichever way the influences worked, and they are beautifully explored in

Carla Gottlieb's *The Window in Art*, windows with glass and the Renaissance seem deeply linked.

The window-like panes of glass had another effect. The major development in perspective in the fifteenth century, which we are told was a vast and unprecedented step, was when people began to conceive of the painting as if it were a pane of glass that cut across sight. Leonardo famously described the transparent cross-section in the cone of sight that bisected vision. In a shorter version he wrote: 'Perspective is nothing else than seeing a place or objects behind a pane of glass, quite transparent, on the surface of which the objects that lie behind the glass are to be drawn. These can be traced in pyramids to the point in the eye, and these pyramids are intersected by the glass plane.' This is such an important idea that it may be worth quoting also his longer account as follows:

> They should understand that, when they draw lines around a surface, and fill the parts they have drawn with colours, their sole object is the representation on this one surface of many different forms of surfaces, just as though this surface which they colour were so transparent and like glass, that the visual pyramid passed right through it from a certain distance and with a certain position of the centric ray and of the light, established at appropriate points nearby in space ... Consequently the viewers of a painted surface appear to be looking at a particular intersection of the pyramid. Therefore, a painting will be the intersection of a visual pyramid at a given distance, with a fixed centre and certain position of lights, represented by art with lines and colours on a given surface.

This was not just a definition of a painting in perspective, it was also a practical technique. Using a pane of glass one could now work out the exact mathematics and angles for correct perspective, as Alberti, Leonardo and others did. Secondly, if help was needed to put this new technique into practice, one could actually paint or sketch on the glass and then transfer the marks on to paper later, with exact measurements. Sometimes Alberti's famous invention, the 'veil' (made of threads), was used, but it was explicitly stated that this was really a sort of window without glass. Leonardo recommended the actual use of a pane of glass in a number of ways. For example,

> Obtain a piece of glass as large as a half sheet of royal folio paper and fasten this securely in front of your eyes ... Next position yourself with your eye at a distance of two-thirds of a *braccio* [an arm's length] from the glass and fix your head with a device so that you cannot move it at all. Then close or cover one eye, and with the brush or a piece of finely ground red chalk mark on the glass what you see beyond it. Then trace it on to paper from the glass ... and paint it if it pleases you, taking good use of aerial perspective.

So the flat pane of glass in the window was essential for both working out solutions to problems in perspective and for helping the weaker brethren do really excellent perspective drawings. Again, one wonders where the developments would have gone if windows had been covered in strong white paper.

So one could argue that while we all see the world in perspective, it is very difficult to work out what we see just by looking at the thing itself. Rather as one cannot look directly at the sun, but look only at its reflection, one has, ironically, to look through the artificial medium of glass in order to see the world as it is, as Alice found when she went through the looking glass. If this is correct, then glass becomes as important in clearing our vision and showing us our world as it really is, in relation to seeing and representing, as it was in the other sciences where the mirror, prism, lens and later the telescope and microscope clarified vision.

In all these cases the human eye, weak and entangled with the interpretation of the brain, cannot see clearly. It cannot see the constituents of light, or how it bends; it cannot see objects below a certain size or beyond a certain distance, though they are there in front of the eyes. At the deepest level, the eye sees the world through a glass darkly, that is an eye is a glass with its own distorting lens and interpretative frame. It is as if all humans had some kind of systematic distortion such as myopia, but one which made it impossible to see, and particularly to represent, the natural world with precision and clarity. Humans normally saw nature symbolically, as a set of signs, not for what it 'really' was, undisturbed by the mind. What glass ironically did was to take away or compensate for the dark glass of human sight and the distortions of the mind, and hence to let in more light.

However gifted the artist, if he looked and painted, he ended up seeing and painting in symbols, like almost all those before Giotto. But if one was forced to follow nature, by painting, copying, photographing so to speak, exactly what was there on a pane of glass, or painting an existing painting done by a mirror, in other words copying rather than painting, miraculously it would be more accurate than a painting of nature itself. Once it was done, one could work out why that should be so. One could establish the artificial rules that would deceive other human eyes into thinking that what was represented on a two-dimensional surface was a three-dimensional mirror or photograph of nature.

If this connection between sophisticated glass technology and the new realist art is correct, it is not surprising to find that in the fifteenth century Italy was at the forefront of the new vision. The Venetian glass industry was a glory of the world; the wealth of the small, competing courts was invested in conspicuous consumption. All one had to add was luck and chance, for example Brunelleschi's accidental discovery of the virtue of the mirror, originally developed for another purpose, to give a picture of what a new building *in situ* might look like. The similar wealth and highly developed glass industries of northern Germany, France and the Netherlands gave the new vision its other home.

Long ago the historian Burckhardt suggested that one central feature of the Renaissance was a new concept of the individual, unique to the west and to the period from about the fourteenth century. While many have subsequently queried the dating, suggesting that the heightened concept of the individual probably went back to the thirteenth and perhaps twelfth centuries, it is usually agreed that this tendency towards stress on the individual grew stronger and reached its peak from the fifteenth century onwards. What seems indisputable is that a great shift took place. Furthermore, this was initially a western European phenomenon, as more recent anthropologists have confirmed. Could this also have been connected to glass?

Of course, it would be absurd to believe that there was a

single reason for this, and it is easy to see the force of many of the explanations put forward to account for a reorientation away from the group and towards the individual. One of the most important of these is religion. At a very general level, it has been suggested that it was the single soul posited by the Judaeo-Christian religious tradition that stressed individualism. Others have suggested that the concepts of sin and individual responsibility are the key to the growing individualism.

The link could work in various ways. These include that exercise of choice and free will emphasised by a religion born out of oppression and founded on Christ's individualistic teaching – for example, to follow him and renounce one's family. The link between choice and individualism is obvious. The individualistic westerner is a sovereign chooser, deciding what to do, what to have, what to be and what to believe. Another link could be through the practice developed by the Catholic Church to deal with sin, namely that form of introspection known as the confessional, which made the individual examine him- or herself and become self-reflective about individual personality. Yet although this seems to be a necessary condition, the variations in time and space do not fit the whole of Christendom; for example, Eastern Orthodox Christianity was far less individualistic. So scholars have pointed to other factors.

Some have suggested that the recovery of classical ideas was the catalyst. Others have argued that the growth of a market economy and particularly of money transactions separated out the individual from the wider group. Money, individualism and the morality of the market economy, it is argued, are all closely related. Yet even adding to this other factors such as the growth of republican government and the middle class in Italy and the Netherlands, the full explanation of one of the greatest transformations in human history still eludes us. This is partly because none of these explanations seems to reach down deeply enough into the psychological realm where the changes occurred. Some extra factor, one feels, was needed – something which would not wholly account for the change in itself, but was a necessary catalyst.

This factor, some historians have argued, was the development of fine glass mirrors, which proliferated at the very time and in the very area where the change took place and could have provided the final factor that was needed by allowing people to see themselves in a new way. Some who have traced the rise of autobiographical writing during the Renaissance have suggested that this 'discovery of the self' was linked to mirrors. Likewise it is pointed out that Renaissance artists such as Dürer explored the inner man through the use of mirrors during their painting. This is an argument forcefully put by Lewis Mumford and he cites the self-examining portraits of Rembrandt as the high point in this artistic introspection.

The timing of the causal link is right; good mirrors developed in almost exact pace with the development of a new individualism between the thirteenth and sixteenth centuries. The geography is right; the epicentres of Renaissance individualism in painting and other art forms were Italy and the Netherlands, two of the most advanced areas of mirror-making and their use. The psychological link is plausible; people saw themselves in a new way that detached them from the crowd and allowed them to inspect themselves more carefully. We can see the process at work in a number of great artists. Yet as with all supposed connections there are doubts. Most cultures have mirrors of some sort and one wants to know more about how mirrors are used, the relative clarity of metal and glass mirrors and so on.

On the question of use, it is clearly important to discover the way in which mirrors were regarded. In the west they were largely looked into to see the person. This was both a cause and consequence of growing individualism. In China and Japan and perhaps other civilisations mirrors were used for different purposes. It is worth examining one example in some detail to see the differences that mirrors and culture could make.

A number of analysts, both foreign and Japanese, agree that in Japan mirrors were traditionally used in a very different way from that in the west. They looked through the mirror image and through the 'observing self'. The mirror was not an instrument of vanity and self-assessment, but of contemplation, as can be seen in Shinto shrines where the mirror is the central object. The individual does not gaze into the mirror to see a rounded portrait of the physical and social person in front of the mirror, but to gaze through the physical into the innermost, mystical self. The Japanese have tried to explain how people in the west look in mirrors out of a compound of narcissism and individualism, while the Japanese look through the mirror. If the Japanese want to see the reflections of their personalities, they do so through the mirror of society, the reflection of the effects of their actions and words on others.

Both the material and the use were thus different. Mirrors were sacred objects in Japan. They were kept in shrines. They also, it would appear, were kept for special use, in particular grooming oneself, but were not hung on the walls of ordinary rooms. The traveller Thunberg noted this in the later eighteenth century: 'Mirrors do not decorate the walls, although they are in general use at the toilet.' Thunberg also noted that among the things not found in people's apartments were 'looking-glasses'. There were no glass mirrors, he said, for all were made of steel. It appears that one reason for this was that the superb workmanship in steel of the Japanese allowed them to make steel mirrors of high quality. In fact, the use of steel or bronze may have limited their size and possibly their effects on the viewer. The fact that they tended to be convex may also be important in limiting the amount that could be seen in them – they were basically for hair arranging, eyebrow plucking and teeth blackening in domestic use.

The Japanese metal mirror reflects back only about 20 per cent of the light that hits it and is slightly coloured. It could thus be argued that because the eye is so sophisticated and appreciates visual perfection, the difference between a very good, silver backed, glass mirror of the kind being made in Europe from the thirteenth or fourteenth century, and a quite good metal mirror is not just one of degree, but of kind. A subtle change in the artefact permits a different perception of life.

Therefore we could say that what we call a 'mirror' in most cultures encouraged imagination and stimulated thought, but not a deep staring at what was portrayed. The western glass mirror showed what appeared to be real even though it was in fact almost magically reversing things, representing three-dimensional space on a flat surface and encouraging the eye to see foreground and background.

Mirrors are indeed extraordinary and it is not too fanciful to believe that the development of the glass mirror in only one civilisation not only altered its art, which can be shown, but also gradually altered the whole perception of what human beings are. One certainly has an 'elective affinity': individualism and high quality mirrors grow together. Yet one cannot see a simple and direct causal link of a necessary and sufficient kind. Glass mirrors, on their own, would not have effected the huge transformation which we call Renaissance individualism. Yet they may have been one of the necessary enabling causes, without which the abstraction of the individual from the group would not have taken the course it did.

So we have three major ways in which glass and an increase in reliable knowledge and representation in the fourteenth to sixteenth centuries may have been linked. One was through the influence of medieval optics and geometry on the perspective art of fifteenth century architects and painters. A second was through the influence of glass, particularly mirrors, windows and panes of glass, on the technology of enchantment and illusion. Thirdly, through the effect of mirrors on concepts and representations of the individual.

We conclude by trying to assess the weight of these links by asking a number of questions. Is it possible to have a reasonable perspective drawing without optical glass (mirrors, lenses, windows) being widespread in a civilisation? The answer is yes, it is possible, though it requires great skill to get far. Is it possible for such a realistic, perspectival way of representing the world to come to dominate a civilisation without much optical glass? All that can be said is that we know of no such case and we can see reasons why it is unlikely. Is the possession of glass optical instruments, mirrors, lenses, prisms, glass panes in profusion bound to lead to realistic, perspectival art? The answer is probably not. There is no necessity. The Islamic case, although they did not have all of these instruments, provides one reason why not. If there is instituted fear of icons, including mirror images, or even a different role for art, then such a development may never occur. Traditional forms of Indian, Chinese and Japanese art have continued to flourish after the importation of western glass technologies and western perspectival art. This leads to the conclusion that many other things as well as glass influence an artistic tradition. Glass is not a sufficient cause.

Yet it might well be argued that it is a necessary one. It is difficult to imagine that the art of the Renaissance could have reached the realism that we find in Van Eyck, Leonardo, Dürer or Rembrandt in a civilisation without glass. Firstly, the geometry and knowledge of optics that laid the foundations for their work, as is so evident from their own writings, would have been missing. This geometry and optics was dependent on medieval European glass-influenced experiments and philosophy. Secondly, the refinements of medieval knowledge from Van Eyck and Brunelleschi onwards frequently required experiments with glass, with mirrors, lenses and flat panes of glass. That is to say the usual cycle of improved reliable knowledge, innovation of better artefacts, the feeding back of these artefacts into further knowledge, could continue. If it had been halted, as it would have been if Europe had been in the largely glassless situation of China and Japan, or the Islamic world after 1400, then it is difficult to see how that vast revolution which we call the Renaissance could have occurred. So the development of glass did not cause the Renaissance directly. Yet without it, something very different would have emerged.

All of this was, of course, a giant accident. High-quality mirrors, excellent flat glass, lenses – these were originally not designed to push a civilisation away from its inherited cosmology, to encourage individualism, the dethronement of God, the disassociation of sensibility, a new and more accurate knowledge of the 'real' world. They were toys of vanity or comfort, just like fine porcelain. So if Europeans had mastered the secret of porcelain, or been able to cover their windows with paper from the mulberry, probably none of this would have happened. There would almost certainly have been no Leonardo, no Renaissance reversal of vision, and no classical scientific revolution. The unintended consequences of that strange, light-bending transparent substance, glass, gave people new eyes to see and what they saw changed the whole world.

It changed our world not only in itself, but also because it was a link in the chain leading up to the classical scientific revolution of the seventeenth century. If the essence of the scientific revolution was precision of detail, accuracy of recording, understanding of the nature and relations between things, and curiosity driving people on to deeper probing through 'experiments' to see how man and nature were intertwined, then all these constituents are present in the Renaissance. The most obvious example is Leonardo da Vinci himself. If one asks whether such a change is likely to make the observation of nature more 'reliable', one could scarcely do better than compare a painting or drawing by Leonardo to a Mughal miniature or a Japanese painting of the fifteenth century.

Leonardo's depictions are struggling to get at the essence

of the underlying laws of nature; they are all 'experiments', as much as Newton or Boyle's experiments. Each of his successful drawings advances anatomy, or physics, or optics. Reciprocally, the urge to try to represent nature as it is, not as it appears deceptively to us, forces Leonardo on into all branches of knowledge. He needs to understand anatomy, botany, geography, hydraulics, mechanics and so on in order to paint properly. If science is the extension of reliable knowledge, then the developments in painting and architecture in the fifteenth century were in many ways as great a step forward as was the more famous scientific revolution of the seventeenth century. Without those developments, it is impossible to envisage the work of Galileo, Hooke, Boyle or Newton. The essential foundation had been laid, but only in western Europe. In other parts of the world the conjuring with glass and mirrors and later with lenses to deceive the human eye into seeing things more clearly did not occur.

5

Glass and Later Science

Nature and Nature's Laws lay hid in Night.
God said, Let Newton Be! *and All was* Light.

Alexander Pope, 'Epitaph. Intended for Sir Isaac Newton in Westminster Abbey'

WE HAVE SEEN that the foundations for the classic scientific revolution had already been established by the later sixteenth century. With the help of glass instruments, the experimental method, precision, pursuit of knowledge as a valued activity, abstraction and framing, the emphasis on sight and many other central characteristics was already present. This chapter will concentrate fairly narrowly on the period when glass instruments of a more powerful and explicitly scientific nature – microscopes, telescopes, barometers, thermometers, vacuum chambers and so on – were manufactured. We can obtain a preliminary glimpse of this increasingly glass-filled scientific world if we look at the work of the man who laid down the charter for the new experimental science, Francis Bacon.

Bacon imagined the kind of laboratories and equipment they needed for the effective interrogation of nature in his *New Atlantis*, written a few years after the invention of the microscope and telescope in the early seventeenth century. He outlined the necessary glass instruments that would be needed to generate reliable knowledge. Speaking through the keeper of 'Solomon's House' he wrote as follows concerning the analysis of light (and foreshadowing Newton's work on optics):

> We have also perspective houses, where we make demonstrations of all lights and radiations; and of all colours; and out of things uncoloured and transparent, we can represent unto you all several colours: not in rainbows, as it is in gems and prisms, but of themselves single. We represent also all multiplications of light, which we carry to great distance; and make so sharp, as to discern small points and lines: also all colourations of light: all delusions and deceits of the sight, in figures, magnitudes, motions, colours: all demonstrations of shadows. We find also divers means yet unknown to you, of producing of light originally from divers bodies.
>
> We make artificial rainbows, halos, and circles about light. We represent also all manner of reflections, refractions, and multiplication of visual beams of objects.

As far as telescopes, spectacles and microscopes were concerned, he explained that

> We procure means of seeing objects afar off; as in the heaven and remote places; and represent things near as far off; and things afar off as near; making feigned distances. We have also helps for the sight, far above spectacles and glasses in use. We also have glasses and means, to see small and minute bodies perfectly and distinctly; as the shapes and colours of small flies and worms, grains, and flaws in gems, which cannot otherwise be seen; observations in urine and blood, not otherwise to be seen.

The last points towards the work of Harvey on the circulation of the blood and work on the causes of disease. It is no surprise that among the statues of great explorers and inventors described in the *New Atlantis* is that of 'the inventor of glass'. Nor is it surprising that many of the experiments which Bacon described elsewhere in his works started with a phrase such as 'take a glass'.

What Bacon fully realised was the degree to which glass had become the essential tool to aid thinking and understanding and for investigating the laws of nature. This connection, so often taken for granted and hence half-forgotten, immediately becomes obvious if we think of the effects of microscopes and telescopes on changing human perceptions of what lies beneath their normal sight ability, or beyond their usual range, the microcosm and the macrocosm. The previously invisible, the teeming micro-organisms which determine human life, became visible. Meanwhile the far distant and tiny objects of the heavens were suddenly brought closer. This transformation of the spatial dimensions of the world is now so familiar to us all that we take it for granted. Only in privileged moments, such as when we showed a group of villagers in a mountain village in Nepal the tiny living objects in their drinking water through a pair of lenses, does one recapture the sense of awe and surprise which many must have felt with the discovery of the microscope and telescope.

Furthermore it reminds us that the effects of powerful glass tools act at many other levels as well. Sight is humankind's strongest sense and by providing new tools with which to see an invisible world of tiny creatures, or to contemplate distant stars invisible to the naked eye, glass not only made possible particular scientific discoveries, but led to a growing confidence in a world of deeper truths to be discovered. It was clear that with this key one could unlock secret treasures of knowledge, below and above the surface of things, destabilise conventional views. The obvious was no longer necessarily true. Hidden connections and buried forces could be analysed.

All this, of which a new conception of space was part, is supplemented by other glass instruments, in particular the retorts, test tubes, vacuum chambers and other isolating instruments. Glass, we know, has two unique properties. Not merely can it be made in a transparent form so that the experimenter can watch what is happening, but it is also, in the case of most elements and chemical compounds, resistant to chemical change: it has the great advantage of remaining neutral to the experiment itself. Its virtues do not end here. It is easy to clean, seal, transform into the desired shape for the experiment, strong enough to make thin apparatus and to withstand the pressure of the atmosphere when a vacuum is created within it. It is resistant to heat and can be used as an insulator. It has a combination of features we do not find in any other material. Where, as Lewis Mumford asked, would the sciences be without the distilling flask, the test-tube, the barometer, the thermometer, the lenses and the slide of the microscope, the electric light, the X-ray tube, the audion, the cathode-ray tube? The role of lenses and prisms in testing light was behind many of the key experiments in physics in the seventeenth century carried out by Galileo, Kepler and Newton. It is therefore no coincidence that the list of those who became lens-grinders overlaps so closely with the great figures associated with the scientific revolution.

Grinding glass to make artefacts is about the most precise craft skill in the world. It is several orders of magnitude more exact than anything else western craftsmen were doing. Is it any coincidence that so many great scientists (Spinoza, Descartes, Hooke, Huygens, Newton, van Leeuwenhoek) were also glass-grinders? Even if not grinding glass themselves, they were well aware, from using precise glass instruments, what a huge difference tiny variations in surface could make, just as mechanical clocks were making the same message obvious in relation to time. Precision, accuracy, exactness, focusing down on the particular problem, all are deeply affected by mirrors, lenses, prisms, spectacles.

Lenses are only part of the optical repertoire opened up by improved glass. Mirrors were also crucial, not only for their practical importance in surveying and navigation but also for their use in other instruments, such as telescopes, and as aids in optical experiments. The development of prisms, to experiment with light, had immense consequences for Kepler, Descartes, Newton and others. Without these glass tools there could have been little deepening of knowledge of the properties and nature of light. And such knowledge, built back into improved lenses, led to new microscopes, telescopes and finally to that great instrument of knowledge or extending of the eyes, the camera. Glass was manifestly one of the very most important materials in the development of science and technology and has remained so ever since.

The actual details of who discovered what, particularly in relation to such instruments as telescopes and microscopes, are not important and are, in any case, a continuing source of dispute. All that we need to state here is that by the first quarter of the seventeenth century, both the microscope and the telescope had been invented. Whatever the precise chronology, it is clear that a fruitful interaction between the various centres of experiment and technology in Italy, the Netherlands, England and elsewhere generated rapid advances in glass technology, and the new knowledge that improved instruments enabled then fed back into further technical advance.

The way in which developments occurred as a result of a network of interacting knowledge and craft centres across the whole of Europe is well illustrated by the history of that key invention, the microscope. This account also shows the cumulative input of many minds into an artefact and the parallel development of glass and reliable knowledge. The story starts in the Low Countries in the early years of the seventeenth century with the invention of the microscope. Initial development was slow, but within a century some of the fine detail of the living world had been uncovered, and enough interest generated to ensure an ongoing process of discovery and exploration. Red blood cells were seen travelling through capillaries, first in Bologna in the mid seventeenth century, and equipment for seeing blood flow in the tails of small fish became part of the accessory kit of microscopes for over a hundred years.

In 1665, Robert Hooke wrote in *Micrographia*, the earliest detailed book on the microscope, 'If you take a very clear piece of Venice glass [a fragment of a broken wine glass] and in a Lamp draw it out into very small hairs or threads, then holding the ends of these threads in a flame, till they melt and run into a small round Globul, or drop, which will hang at the end of the thread ...' Hooke was describing the manufacture of the first high-power microscope lenses, which by 1683 would enable Anthony van

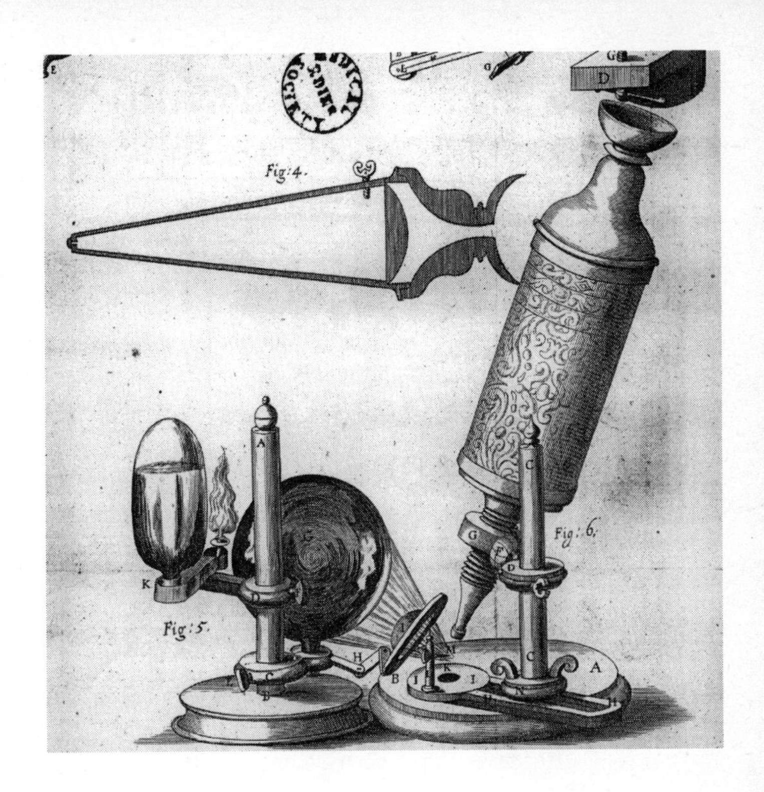

7. Micrographia

A compound microscope by Robert Hooke, from Hooke's *Micrographia*, London, 1665. Used by Hooke to produce one of the first illustrated books of microscopic objects, it helped to inaugurate the discoveries which, for instance, led to an understanding of infectious diseases. The microscope revealed a world invisible to the naked eye, which only science (with the help of glass) could probe.

Leeuwenhoek, a Dutch draper and microscopist, to obtain the first glimpse of bacteria, thus setting in train the long line of research that led, in the nineteenth century, to an understanding and partial conquering of infectious diseases.

The process of improvement of the microscope was driven almost entirely by curiosity, for there was no economic use for it until 1840. Early microscopes produced quite indistinct images, with coloured fringes around the objects and fuzziness at the edge of the picture. It took a period of two hundred and forty years to improve the microscope to a level of refinement at which it would reveal details of bacteria and of the mechanism of dividing cells and reproduction. Crucial contributions came from all around Europe. Great improvements in the glass for the lenses were made by Pierre Guinand, a Swiss working with Joseph Fraunhofer at Benediktbeuern, and by Otto Schott in the Carl Zeiss Company at Jena. To correct the coloured images produced by simple lenses, which limit magnification and clarity, two differently shaped lenses made of two different types of glass were needed. Newton tried to eliminate colour fringes by combining lenses of different shapes but made of the same type of glass and did not succeed, concluding that the task was impossible. Before 1670 only traditional glass had been available – but then a second type, lead glass, was developed by Italian glass-workers working in London for the English entrepreneur George Ravenscroft.

He succeeded in producing this new type of glass, but had no thought of its use in telescopes or microscopes. Ravenscroft had developed it for use in wine glasses, jugs and bowls. It was seventy years before John Dolland in London used it in telescopes. Dolland was the son of a French Huguenot refugee who came to England after the revocation of the Edict of Nantes in 1685, just one hundred years after Flemish opticians fled Antwerp and established the optical industry in the United Provinces which subsequently developed the practical form of the telescope.

The theoretical development of the new achromatic lens, central to the improvement of the microscope, had important inputs from Klingenstierna, Professor of Mathematics at Uppsala, and from Leonard Euler, a Swiss mathematician working at St Petersburg. Joseph Lister, a London wine merchant, encouraged by the Scottish scientist David Brewster, made further improvements to the theory of the microscope.

Ernst Abbe, a German physicist working in the Zeiss company with Schott, the glass-maker, further refined the optical theory and the practices of manufacture to a standard at which the microscope could open up a new understanding of the world.

The microscope was crucial to Pasteur, the French chemist and microbiologist, in his work on the germ theory of disease. At the time it was widely believed that life, in the form of flies, maggots, moulds and so on, could spontaneously generate from putrefying matter, a view that had to be solidly disproven before the causes of infectious diseases could be properly investigated.

Pasteur undertook a series of experiments using a broth of yeast extract and sugar in little swan-necked glass flasks. After boiling the broth for several minutes to kill any organisms that might exist in it, he sealed off the top of the neck on some flasks, allowed heat-sterilised air to enter other flasks, and filtered out airborne microbes with cotton wool in yet other flasks. After being left for two or three days in a slightly warm oven, flasks that had been left completely sealed or

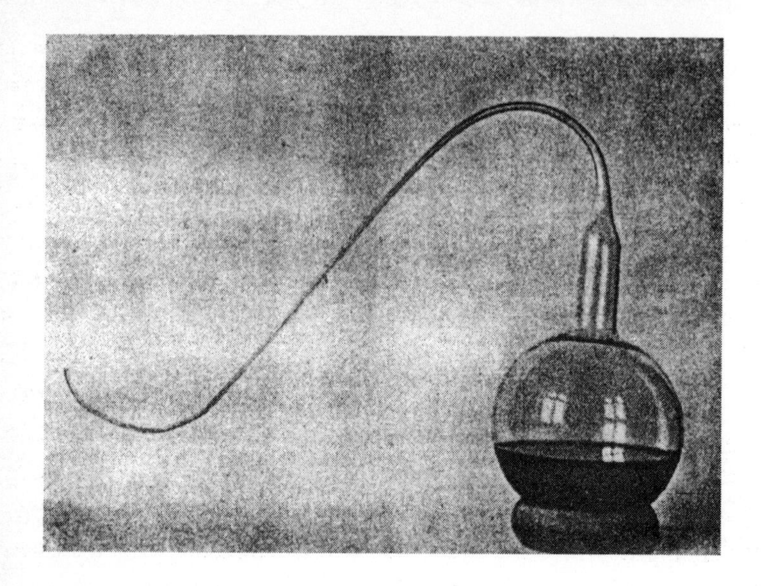

8. Pasteur's bottle

Even as late as the 1860s, when this illustration was made, the possibility that life could spontaneously generate and hence that disease might arise without outside intervention from dead organic matter was still widely believed. Pasteur disproved spontaneous generation in a series of experiments using many swan-necked glass flasks such as this one (see Appendix two for further details).

given access only to sterile air remained free from mould and putrefaction. All other flasks developed a variety of life forms on the surface of the broth. It is difficult to conceive of any material other than clear glass which could have been used for these simple but conclusive experiments. This was an essential stage in the understanding and combating of infectious disease, since infection would be a far more intractable problem if microbes could appear spontaneously out of dead matter.

Later, the microscope would be the foundation of an even greater breakthrough. Our understanding of genetics springs from the discovery of DNA and the double helix, which in turn rests on the discovery of the chromosome and the processes of cell division. This relied on the steady improvement in the resolving power (the ability to see fine detail) of the microscope.

The story as related above is misleading in its simplicity. Many more individuals were involved, each situated in a complex network essential to their contribution. Nevertheless the tale is typical of any of the stories of product development, that is the interaction between ideas and things over long periods of time, which collectively make up our modern world.

Although it is not too difficult to see a general link between scientific tools of glass and increased reliable knowledge, the subtler connections are so much more complex that they often escape us. In most major scientific discoveries there is a long chain of causation which, if broken at one point, cannot be completed. Often we find that glass in some form, whether as container or lens, is one vital link. Thus while it is not the proximate cause, it is an essential one.

We have looked carefully at the twenty 'great scientific experiments' selected and described by Rom Harré, the Oxford historian of science, as 'twenty experiments that changed our view of the world'. They stretch from Aristotle's work on the embryology of the chick through to work in the twentieth century on quantum mechanics, imprinting, genetics and perception. It is clear that glass instruments were essential for twelve out of twenty of the experiments. Most of the other eight required knowledge from earlier experiments which had used glass. (See Appendix 2 for details.)

In order to see the complex way in which glass can be an essential link in an apparently unconnected but important development, let us look at just one intriguing example, the relationship between glass and the history of the steam engine. At first sight the connection is not obvious. Steam engines do not seem to have any glass in their manufacture, so why would it have been impossible to build a steam engine without widespread use of scientific glass?

There can be no doubt of the importance of the steam engine. It is the symbolic and actual tool of the change from an agrarian to an industrial world. This device transformed the speed of travel, the speed of weaving, the effectiveness of mining, the distribution of clean water and many other things. It allowed civilisations to move from a situation of limited energy from plants and animals, to one in which they have access to the massive deposits of millions of years of carbon energy. It was made into an effective invention in north-western Europe as the result of a chain of causation that ran directly through glass.

In fact, the steam engine was only one of the converters

which has enabled mankind to release absorbed sunlight and accelerate matter, later giving way to internal combustion engines, gas turbines and so on. All these devices worked by deriving useful mechanical work from expanding gas. An understanding of the properties of gases, the gas laws, the relationship between the volume of a gas and its pressure and temperature, was advanced greatly during the seventeenth century. This was achieved in part by experiments with the newly invented air pump in Germany, England and the Netherlands, following the invention of the barometer, and the demonstration of the existence of the vacuum in Italy in the 1640s.

Glass was not required in the manufacture of steam engines of the 'atmospheric' type, the first generation of engine, in which the weight or pressure of the atmosphere was used to press down a large diameter piston into the partial vacuum created under the piston by the condensation of steam. However, the designers of the first engines needed a good understanding of the nature of the atmosphere, in other words that the air had 'weight' and could exert pressure, that it was possible to make a closed vessel with much less air in it than existed in an open vessel, that steam could be condensed to achieve such a situation and that the resultant pressure difference could be used to exert considerable force. This set of understandings, perceived only dimly if at all in 1600, had been deeply investigated and even quantified in Europe by 1700 through a series of carefully executed experiments, in several of which glass was crucial.

In the 1640s Berti in Rome, and a little later Torricelli in Florence, performed experiments which strongly suggested the existence of space empty of matter – what we would call a vacuum. Each performed the experiment with a long vertical tube initially filled with a fluid. Berti's tube was made of lead, around thirty feet tall and supported on the side of his house. On the top was cemented a flask made of glass which contained water. Torricelli, using mercury, used a much shorter tube, made entirely of glass and sealed at the top end. The tube would have been around three feet long. In each case the fluid in the tube was allowed to flow out at the bottom, leaving a vacuum at the top. The nature of the vacuum and the reason for a constant and fairly reproducible height of remaining fluid stimulated a huge amount of subsequent experimental work.

Torricelli's experiment provides an interesting case of how innovation occurs. In the absence of glass, Torricelli could not, realistically, have said, 'I need a material which I can see through, in the form of a tube of indeterminate length' (for to arrive at the correct length already required a number of exploratory experiments). 'I need something that is closed at one end and strong enough to support itself and a heavy column of mercury.'

If we consider the next stage, we can assume the existence in Italy in the 1640s of a competent glass industry used to producing clear glass and making blown bottles or decorative ware or blowing large bubbles of glass as a stage in the production of sheet glass. So Torricelli could pay a glass-worker to draw out one of his large bubbles of glass into a long tube. This would be an innovation, but one well within the capabilities of the time.

The final stage of manufacture, the closing of one end of the tube, sometimes with a simple melted seal, sometimes with a glass bubble blown at the melted end, could have been performed either by the skilled glass-worker or by Torricelli or an assistant. Thus, the process is a mixture of selecting from what was already available and explicitly designing new things. A similar story could be told about Robert Boyle, whose glass vacuum pump chamber could have been made only in a few workshops in the world because only in them could a sufficiently large round glass object be made.

This shows the way in which if you move back along the chain of causation, you find that you cannot have barometers or air pumps and their chambers and gas laws without clear glass, as Robert Boyle knew so well. If you use metal, pottery or porcelain containers or tubes you simply cannot see what is going on. Thus glass was an essential link in the chain which lay behind the major power sources of the industrial revolution, for it provided more accurate knowledge of the laws of nature.

That glass was the essential ingredient for most of the significant tools of scientific investigation in the seventeenth century is obvious. Consider for example Knowles Middleton's comment that 'The invention of six valuable scientific instruments within a few decades of the seventeenth century undoubtedly had an immense effect in accelerating the development of science. These instruments were the telescope, the microscope, the air pump, the pendulum clock, the thermometer and the barometer. They all made experiments and measurements possible which had been unimaginable before ...' All but the pendulum clock depended on the availability of high-quality clear glass.

Three final examples of the indirect effects of glass may be given. One concerns the ubiquitous, necessary, but normally overlooked gadget of modern life, the humble light bulb. At the beginning of the third millennium we live in a world of artificially created light. The burning of waxes and oils had stood humanity in good stead for some thousands of years in order to extend daylight. With the advent of electricity early in the nineteenth century the possibility arose of large numbers of small, economical, practical light-producing devices. The problem, which was confronted by a number of experimenters, most notably Thomas Edison, was how to produce a small loop or wire of material which would conduct electricity and glow white hot, thus emitting light. It is quite easy to do this with a thin wire of platinum, for instance. It is quite another matter to produce a device which will go on working for weeks or months. Placing the filament in a glass container and evacuating practically all of the air with newly developed high-vacuum pumps solved the problem. Eventually it was found that the vacuum could be replaced with carefully purified inert gases. The clear glass container has remained an essential necessary component of both the filament lamp and the more modern fluorescent lamp to this day.

A second example comes from the art of navigation, without which the trading empires which provided the foundation of modern global economies could not have been successfully established. Until the early eighteenth century, long-distance trade and navigation were extremely hazardous and hardly profitable because of inaccuracies in measuring latitude and longitude. Numerous expeditions were lost and expensive fleets destroyed through gross errors of reckoning. Ships found themselves in the middle of the Pacific without any knowledge of which way to proceed, and even voyages along the Atlantic coast of Europe or to America were fraught with extra danger because, once out of sight of land, it was almost impossible to estimate accurately where one was. In relation to latitude, the problem was slightly easier. It had for long been possible to do moderately precise measurements of the angle of the midday sun using such devices as the backstaff and other instruments. In the early eighteenth century the accuracy of these measurements was increased significantly by the use of the sextant, a device which cannot be made without glass. The sextant uses both half-silvered and clear glass. The silvered glass is used to reflect an image of the sun and by adjusting the image you bring it down on to the horizon which you see through the clear sighting telescope. This instrument, if used carefully, can enable you to determine your latitude to within ten or twenty miles.

More difficult was the establishing of longitude. It was the development of the precision clock or chronometer, sufficiently robust to withstand the salt spray in the air, the large variations of temperature and humidity, and the constant movement encountered on a long sea voyage, where storms and gales were frequent, which provided the first really practical method of establishing longitude. Glass is not used in the mechanism of the chronometer, but is an essential component without which it would have been impossible to make a usable, protected device to carry on ships. The glass on the front meant that the workings were sealed off, yet visible.

A number of highly significant navigational aids which became increasingly important during the eighteenth century were also dependent on glass. Telescopes and binoculars were widely used to see ahead. Lighthouses, harbour lights and boat lanterns to show port and starboard were essential for navigation in difficult and crowded waters. Inside the ship, lanterns protected by glass made working at night feasible where an unprotected candle would soon have been extinguished.

9. Harrison's chronometer

A front view of H.4, Harrison's prize-winning watch, which played a critical role in the solution of how to determine longitude accurately while at sea. This and the other miniature chronometers would have been useless without the glass front which protected them against damage from various hazards at sea – the wind, salt in the air, the danger of falling objects during storms.

The third example lies in the field of representation of visual images projected over space and time. The earliest experiments with projecting images through the use of the *camera obscura* are very old and did not require glass. The simplest of these devices was merely a small hole in the shutter of a window, projecting an (inverted) image of the outside world into a darkened room. The early experiments out of which photography developed in the first half of the nineteenth century required glass at several stages, in the lens of the camera, the photographic plate and the dark-room. Without glass there would have been no photography and everything that flows from it up to our photograph-soaked world. Furthermore, obviously, without photography there would have been no development of moving films from the last decade of the nineteenth century. Although television uses a different technology from film, it still depends on cameras and screens which have glass as an essential component and it seems pretty unlikely that it would ever have developed without the inspiration of moving films in the preceding years. Gas or steam turbines to this day generate the power electricity, which makes television possible, and neither of those would have existed without the development of the steam engine and understanding of gas laws. So our world would have been without photography, moving pictures and television but for the presence of glass. The influence on such things as the generation of desire, the rapid transmission of information over long distances, the preservation of information about the past is obvious once you consider what a world without these information technologies would be like, not to mention computers which, at first at least, required glass screens.

Putting it simply, none of the following sciences would have existed without glass instruments: histology, pathology, protozoology, bacteriology, molecular biology. Astronomy, the more general biological sciences, physics, mineralogy, engineering, palaeontology, sedimentology, vulcanology and geology would also have been very different. They are all disciplines which rest firmly on the availability of clear glass and the skills to manipulate it. They are good examples of chains of causation in which glass is a crucial link.

So we can conclude by noting how much of our world would not now exist without this remarkable material. Without clear glass we would not have had gas laws, no steam engine, no internal combustion engine. Without clear glass we would not have had the visualisation of bacteria, no understanding of infectious diseases and the medical revolution since Pasteur and Koch. Without the chemistry which depended crucially on glass instruments we would have had no recognition of nitrogen and so no artificial nitrogenous fertilisers and hence much of the agricultural advance of the nineteenth century onwards would have been lost. Without clear glass there could have been no telescopes. Astronomy would be limited to visual observation. There would have been no knowledge of the moons of Jupiter, no obvious way to prove that Copernicus and Galileo were right. Without glass we would have had no understanding of cell division (or of cells) and thus no microbiology and no detailed understanding of genetics, certainly no discovery of DNA. Also, of course, without spectacles, a majority of the population (in the west at least) over the age of fifty would have very great difficulties in reading this or any other book.

6

Glass in the East

A man that looks on glasse,
On it may stay his eye;
Or if he pleaseth, through it passe,
And then the heav'n espie.

George Herbert, 'The Elixir'

HOW CAN ONE TEST the thesis that glass was a necessary, if not sufficient, cause of the explosion in reliable knowledge in western Eurasia? Since we cannot perform the experiment of rerunning the history of western Europe with glass subtracted, the only method left to us is the comparative one. What happened in other civilisations, and in particular how do the two halves of Eurasia compare? This is not, of course, conclusive. It can be used as a negative test. One can plausibly argue that if one could find a case (for example, India, China or Japan) where the kind of growth in reliable knowledge we have found in western Europe occurred in a civilisation where glass was little used, then it would disprove the hypothesis that glass was a necessary condition for the development of 'science'. On the other hand, if India, China or Japan do not have that expansion of reliable knowledge and also do not have much glass, it might still be the result of factors other than its absence.

Glass can never be other than one of a number of necessary causes, and never sufficient in itself. Hunter-gatherers do not primarily lack science because they lack glass. The Romans made wonderful glass but never developed what we would now term 'science'. Nevertheless, it is still useful to create a control study by looking at other cases. We shall move eastwards from the Mediterranean, briefly surveying the histories of Islamic, Indian, Chinese and Japanese glass.

In some ways, one of the most instructive histories of glass is that of its fate in Islamic civilisations. After the collapse of the Roman Empire, the centre of glass-making shifted back to the eastern Mediterranean, the area where glass had first been discovered and developed – Syria, Egypt, Iran and Iraq. For a while the Sassanian Empire which ruled wide areas of western Asia from AD 224 to 651 held this region. The glass here was usually pale green or transparent. Techniques included blowing, casting and pressing, wheel cutting, with stamped and applied decoration. Many of the objects were very beautiful and travelled long distances; for example, two of them were found in Japanese tombs of the sixth and eighth centuries AD.

In the seventh century, as part of the spread of Islamic civilisation, the Arabs destroyed the Sassanian Empire but the glass industry was not extinguished. Hence, since Islam had also absorbed two other great areas of glass-making, the Syro-Palestinian and the Egyptian, the new civilisation was heir to many of the most advanced techniques of glass-making. Yet, for about a hundred years after the invasions glass-making declined; it began to revive about AD 750 with more settled conditions particularly in Baghdad. By the ninth century a distinctively Islamic style had been established, which was famous for its exquisite craftsmanship. Glass-making was very widespread and varied from useful vessels to exquisite luxury ware.

Those within the Islamic area created wonderful decorated glass hardly surpassed before or after. Lustre-painting, skilful engraving, enamelling, gilding – in all of these the Islamic glass-makers excelled. Very large quantities of bottles, bowls, jugs, gaming pieces, small discs for establishing the weight of coins, lamps (particularly for mosques) and mosaic glass were produced. This glass was traded all over Eurasia and became as important in the vast New World empire of Islam as it had been in the Roman Empire. It has been found in places as far apart as Scandinavia, Russia, East Africa and even China. It was in Syria during the thirteenth and fourteenth centuries that the most glorious glass was produced. In Aleppo and then Damascus glass was created on a wide scale, including brilliant examples decorated with animals, birds, flowers and arabesque foliage. Among the objects there were sprinklers, globes, footed bowls, beakers and long-necked bottles. Mosque lamps (really lanterns) for use in schools and mosques were particularly beautiful and important, the nearest equivalent in their symbolism and widespread use to the growth of stained glass in western European churches. They illustrated the words of the Koran, 'Allah is the Light of the Heavens and the earth. The likeness of His light is as a niche, in which is a lamp. The lamp in a glass, the glass as it were a shining star.'

Unfortunately we know little about the use of glass for functional purposes. Yet we do know that there was a wide production for scientific instruments, for alembics (distilling) and cupping glasses for bleeding. Official weights and measures were also made of glass. It seems likely that it was also used in a limited way to make mirrors, plano-convex lenses for magnification and for other purposes. The fact that Arabic thinkers at this precise time revolutionised mathematics, geometry, optics and chemistry is surely not a coincidence.

Another major category was glass to hold perfumes and cosmetics, namely small flasks and other containers for ointments and unguents. This is interesting and is far less developed in western glass-making since Roman times. A third group namely tableware, is the largest, consisting of bottles, bowls and dishes. One thing we would like to know about is the development of drinking glasses. Is it true that the ban in Islam on the consumption of wine had an effect? The question is prompted by the importance of wine-glass manufacture in Italy on the development of fine glass.

Finally, it appears that there was only a very limited development of window glass, which was hardly used in the traditional houses of the Middle East where it was important that air circulated in the hot season. Books on Muslim architecture occasionally mention the use of small stained glass segments in religious and secular buildings, but there is little evidence beyond this. This absence of the development of flat glass is important; much of what is most extraordinary, particularly in northern Europe, arises out of an intersection of climate, Christianity and glass in the development of stained glass in churches, and plain glass windows from Roman times onwards.

The rapid collapse of the glass industries within Islam is a mystery. The simple explanation that Mongol invaders effectively wiped out glass, clearly had a good deal to do with it. The first wave of destruction occurred in the twelfth and thirteenth centuries and encompassed the northern half of Islam and parts of Russia. Mongol invasions had destroyed the flourishing glass industry in Kievan Russia in the early twelfth century. The flourishing Persian industry was nearly all destroyed by Genghis Khan in the early thirteenth century.

The second wave of destruction occurred in the fourteenth century. The destruction of glasshouses and the deportation of the glass-makers from Damascus by Timur in 1400 more or less put an end to the golden period of Islamic glass-making, but the decline in quality and quantity had, in fact, started about fifty years before Timur's destruction which suggests other forces were also at work. Little glass of any quality was produced in Syria or the neighbouring regions after 1400. The glass industry had simply died out and it was Venice that began to fulfil the need for luxury glass. Whether the end of glass-making in the fifteenth century was the result of competition from the west, the spread of plague in the cities, the deportation of workers or something else is still unclear. All we know is that high-quality glass ceased to be made anywhere in the Islamic world for some centuries after 1400.

This is a rather extraordinary story. In the period between about 700 and 1400 the leading glass region in the world was within Islamic civilisation. It was also the leading area for medicine, chemistry, mathematics and optics (physics). Then, just at the point when European glass was transforming science and vision, glass more or less disappeared from Islam. Surely these are not just coincidences, but some kind of elective affinity, at the least, is involved. But what may have made the European development from about 1200 onwards so much more powerful in the end, is that the thinking tools of glass – particularly lenses and prisms, spectacles and mirrors – were emphasised in a way that, at present at least, does not seem to be the case in Islamic glass-making. Double-sided lenses and spectacles, flat planes of glass (as used in Renaissance painting) and very fine mirrors (as produced in Venice) were never developed in the medieval glass traditions of Islam. Is this the crucial difference?

The story after 1400 is quite briefly told. A little glass was produced in Turkey under the Ottomans, but glass techniques had to be reintroduced from Venice in the later eighteenth century. There is evidence of a little glass made in Turkey in the sixteenth century and in Iran in the seventeenth century, but it was of low quality. There are other instances of minor glass manufacture, but in general there is almost no authenticated glass manufactured in the Middle East between the end of the fourteenth century and the nineteenth century.

It is tempting to speculate what would have happened if the Mongols had not smashed first the glass-making of the north and then that of the south. If, after 1400, Venice had vied with a vigorous Islamic glass industry making fine mirrors and lenses, our world might now be rather different.

The fate of glass in India is equally interesting. Here was a vast and sophisticated continent which excelled in many technical processes over the ages, in iron and pottery, in weaving and spinning, in woodwork and basketwork. It was situated adjacent to the area where glass was first developed (Persia and the Middle East generally) and had constant trade relations with that area. If there is something inevitable about the progress of this technology, we might well have expected glass manufacture to have blossomed in India. What then can we learn about its history?

In the several thousand years before the birth of Christ there appears to have been a widespread knowledge of glass, but its use was mainly for decoration. Early Indian craftsmen made glass beads and bangles, ear ornaments, seals and discs. The knowledge and technology were there, and Pliny stated that no glass could be compared to the Indian, though it is worth noting that he gives as a reason that it was made from broken crystal. There seems to have been a surge in the making and use of glass in the period from about the birth of Christ to the fifth century and some have claimed that glass became widely used. Foreign glass objects, including wine-glasses, were also being imported and it is clear that the technique of blowing glass was known in India. At this point it looked as if India was moving in the same direction as the lands to the west.

Yet the industry then faded away. From the golden period of the Guptas, from about AD 450, the glass industry in India declined to a point where it hardly existed a thousand years later. Only a few bangles and glazed bowls from this period have been recovered. In the Bahmani Period (1435–1518) there was a small revival. Layered glass bangles and beads and bowls dating from then were found throughout the Deccan. Yet when we compare this to what had by now happened in Europe, we notice the conspicuous absence of windows, mirrors, lenses, spectacles and the widespread use of glass for drinking vessels.

In the Mughal period, Persian craftsmen were brought to court and glass was manufactured. Clear glass was uncommon; usually it was of a deep copper blue and ornamented with flowers and other decorations. Hukkah bowls ('hubble bubble') were decorated with glass and some bowls and spittoons of glass were made. By a curious twist, while glass began to be used for mirrors, it was used on the *back* of a metal mirror as decoration (usually green or light brown in imitation of jade). Glass was thus used for fine luxury items for the nobility. Most of what remains from this period dates from the later seventeenth century.

The divergent development of India when compared to western Europe could be seen as the impact of the Portuguese and British traders began to be felt. Spectacles of glass from the first quarter of the sixteenth century have been found, and there is a great deal of evidence of the import of spectacles by the East India Company from the early seventeenth century. The correspondence of this company indicates that looking glasses, spectacles and other glass objects were in demand. Dutch bottles for rose water, gin, ink and so on were also now widespread. Imported glass seems to have been quite widely used, but it is difficult to know how much native production there was even in this period. Certainly by the later eighteenth century, there are interesting descriptions of native glass kilns. In the nineteenth century there was quite a large indigenous glass industry, though it seems that the main products were bangles and vessels and little evidence is to be found of the making of mirrors, window panes, spectacles, lenses, etc.

The quality of Indian glass was a problem during this

period; it was full of impurities. This had various consequences. The lack of transparency led to the demand for superior foreign glass, and particularly from the later seventeenth century for English lead glass, so that the local industry was almost crushed.

The puzzle is this. While Indians had a full knowledge of all the techniques of glass-making, they hardly developed an industry, even before foreign competition could have killed it off. Historians have suggested several contributory factors. One concerns the materials for glass. Some suggest that there was a shortage of natron, natural alkali, in India. This may be important, but if other factors had been propitious one suspects that, given the widespread cottage industry of the nineteenth century, this obstacle could have been overcome.

Two other interrelated causes for the slow development of glass have been suggested. One was the low position of glass-makers. As with all those who turn nature into culture (black-smiths, tailors and leather-workers), glass-makers were relegated to the bottom of the caste system. Thus glass-making would not attract educated or wealthy people. This was also linked to social snobbery and religious restrictions. It appears that glass was not a high-status object which the rich and sophisticated coveted. Religious texts suggest that glass was regarded with some contempt. Certainly it does seem to be the case that its main use was to try to imitate something else – jade and precious stones, china and porcelain. It does not seem to have been valued for itself. This is a circular process. As glass is more highly valued, money is spent on it and on developing finer glass, which in its turn becomes more attractive. This was the pattern in the west. If it is regarded as a second-rate alternative to other things, impure glass will do and its uses are restricted. It attracts less and less.

If we take a wider view, various things stand out from this story. Firstly, the non-development was not the result of either lack of knowledge or lack of craft skills. Both were in as great abundance in this area as around the Mediterranean where glass developed so rapidly. Secondly, India is a prime example of a civilisation which over a thousand years almost forgot about glass. Having been quite widespread by AD 400, at least for small decorative items, a thousand years later it had almost disappeared. Thirdly, it is not difficult to see functional reasons for this, quite apart from the materials side. If we examine each of its major uses, we can see why India did not need glass. Firstly, it had an ancient and very widespread pottery tradition. Cheap pots and drinking vessels dealt with the storage and drinking needs better than the costly glass vessels. Secondly, its climate did not make glass windows a high priority, so flat glass would not develop. Thirdly, it had plenty of good brass and other metals for mirror-making. It may therefore not be necessary to invoke Hindu or Islamic attitudes to glass in the explanation of why India remained, essentially, a civilisation which did not develop glass.

Yet the consequences were incalculable. Among these are the possible effects on Indian science. It is well known that India was very advanced in its mathematics, giving the west the concept and sign of the zero, for example. Yet after the period up to about AD 500 the mathematics became more and more abstract and pure. Nor was there much development, as far as we know, of geometry or optics. The practical experiments and testing of mathematics which glass allows through the use of mirrors and lenses were not possible in India. Secondly, there are the effects on Indian art. As we have suggested, glass is one of the crucial features which influenced the revolution in western art, with perspective, depth and realism. The fact that Indian art, from the medieval period through the famous Persian art of the Mughals, remained two-dimensional and symbolic, was perhaps also influenced by the absence of glass. Thirdly, the concepts of the person and individual were deeply affected, particularly by the absence of glass mirrors, as we shall see.

For most of history China was technologically the most sophisticated of civilisations and so we may wonder what the Chinese made of the extraordinary substance we call glass. From the initial perspective of the west, the career of glass in China over the last three thousand years is puzzling. A civilisation that produced some of the most creative craftsmen in history, excelling in pottery, metal-working, print-making and weaving, contributed almost nothing in the field of the development of glass.

By about the sixth century BC glass was probably manufactured quite widely. The art of glass-casting was mastered in the Han period (206 BC – AD 220) when ritual objects and jewellery were made. The next major turning point was the introduction of glass-blowing techniques, about half a millennium after they had been developed in the Middle East. At first blown-glass objects were imported. But from the fifth century, native glass-blowing was being undertaken.

During the next thousand years there was a mixture of some native manufacture and a good deal of importation of first Roman and later Islamic and European glass. The pieces made and imported were mainly small ritual objects and later toys and other devices, including screens of glass behind which objects moved. It seems that some native glass-making continued. Yet, on the whole, in the thousand years after the introduction of glass-blowing there seems to have been hardly any real development of the glass industry and nothing much has been recovered except small reliquary bottles for religious purposes and some imitations of precious stones. The art was very localised and sporadic, with no long-term evolution.

One way of explaining this is to look at the functions of glass and the attitude towards it. Basically it was seen very largely as an inferior substitute for precious and scarce substances, not as a wonderful material in its own right. Its principal attraction was as a way to imitate cheaply more precious substances such as mineral turquoise. The status of glass and of glass-makers was similar to that in India. Although we are forced to use the word 'glass' for comparative purposes, the substance did not carry the whole load of meanings which we attribute to it. It was just a rather inferior material, less interesting than clay, bamboo, paper and many others.

The second potential use of glass is for containers and vessels of various kinds. Here one might ask what glass could do that fine porcelain could not. Writing of the later seventeenth century, the great Jesuit historian, Du Halde, made a comparison of porcelain and glass and thereby gave an important insight into one of the main reasons for the absence of glass in China.

> They are almost as curious in China, with respect to Glasses and Crystals that come from Europe, as the Europeans are with regard to China-ware; and yet this has never induc'd the Chinese to cross the Seas in quest of it, because

> they find their own Ware more useful; for it will bear hot Liquor, and you may hold a Dish of boiling Tea without burning yourself, when you take it after their way, which you could not do even with a Silver Dish of the same Thickness and Figure; besides China-ware has its Lustre as well as Glass, and if it is less transparent it is likewise less brittle.

He then goes on to show that porcelain, like glass, can be cut with a diamond to make patterns. Thus Du Halde stated the obvious fact that one is not likely to need glass for hot drinks when one has chinaware. The role of ordinary pottery is also important. China is, along with Japan, one of the great potting nations and pottery has many advantages. Pottery is much cheaper and holds hot liquids very well. A tea-drinking nation is unlikely to develop the same kind of wonderful wine-glasses as the heirs to Roman glass.

Moving to windows, it is obvious that with good oiled paper and a warmer climate, certainly in the south, the pressure to make glass windows was largely absent in China. This is part of a much wider set of differences which are beginning to become obvious. For example, southern Chinese architecture largely consists of buildings made of woodwork and lattice – much more like light tents than buildings. Hence it would have been more difficult to place glass windows in these frail, non-weight-bearing walls. The houses of the Chinese peasantry were not suitable for glass windows, even if they could have afforded them, and were lit by empty gaps or paper or shell windows. Furthermore, grand religious or secular buildings, built out of stone to last for centuries, hardly existed in China. The equivalent of the cathedrals or noble houses of the west were absent.

China was largely a country with a very rudimentary glass technology until at least the 1670s. Thus what might be termed the uses of glass as tools of living – drinking, storage, decoration of the body, decoration and improvement of the house – were very different, or non-existent. What is crucial to our argument is the way in which the absence of the development of glass technology in these spheres so influenced the development of tools of thought made with glass.

The Chinese had a passion for mirrors, but it was largely for highly polished bronze mirrors, which were often regarded as having magical qualities. Such bronze mirrors could be made into plane, convex and concave shapes and some experiments were made using them as 'burning glasses'. Some have suggested that bi-convex lenses of glass may also have been made early, but these died out by the twelfth century, if they had ever existed. There have also been inconclusive arguments concerning whether spectacles were developed using glass.

In China, glass technologies, the making of coloured and plain glass, glass-blowing, the use of lead and barium, were all known before AD 800. Yet there was little interest in glass from then on until the brief burst of enthusiasm under the impetus of the Jesuits between about 1670 and 1760, which again faded away for a century or so. Thus over much of the period between about 800 and 1650, the precise time of the rush of glass technology first in Islam and then in western Europe, glass technology hardly developed in China.

It is clear that the Japanese knew how to manufacture glass

very early, and could make both coloured and colourless varieties. Dorothy Blair, the leading expert of the history of Japanese glass describes how glass beads and discs were found, and possibly made, in Japan in the Yayoi period (c. 300 BC to AD 300). The technology expanded and use, mostly for beads, increased during the Kofun period (c. 300–710). After the introduction of Buddhism to Japan in 538, glass reliquaries were made. Later, relics were placed in small glass jars. New techniques for bead-making were developed and possibly for making transparent green glass urns.

In the Nara period (710–94) glass-making advanced even further. Many temples had their own glass construction bureaus. Large stores of beads and chests of broken fragments have been found; there are also 'fish-form tallies', and many kinds of cast and coiled beads. Glass-blowing became common and an indication of the quantity of glass is shown by a monument to the emperor Shomu (d. 756), which contained thousands of glass beads, glass pieces, insets, sash accessories and rods for scrolls. In the Heian period (794–1185) glass-making declined, yet there were still some exquisite and complex examples of beads and inlaid glass.

It seems clear that in the early period glass had a spiritual significance for the Japanese, but the range of uses was small. The glass recorded here was for beads, decorations and religious artefacts; there is no mention of windows, drinking vessels or looking glasses (mirrors). Already there is a divergence from the west, for all these other uses of glass had been developed by the Romans or their successors in the west by the twelfth century. We can begin to see a decline in the use of glass occurring in Japan, from about the ninth or tenth century onwards.

While glass beads were still desired, local glass-making

had decreased greatly by the Kamakura period (1185–1333). There were still some works, but glass vessels were probably imports from China. The decline accelerated through the Muromachi period (1333–1568) to a point where glass-making was almost extinct. Even the use of glass beads almost vanished under the influence of Zen Buddhism, which frowned on image worship. The situation was such that by the middle of the sixteenth century the use of glass, except in the making of a few beads, was apparently unknown. Glass was not being made in the Azuchi-Momoyama period (1568–1600). Indeed, the art had been so completely forgotten that the first blown-glass objects brought by western traders and Jesuit missionaries in the later sixteenth century were thought to be made of a new and exotic substance which the westerners must have dug up out of the ground. This is indeed an extraordinary story, though not dissimilar to that of her giant neighbour China. Between about the tenth and sixteenth centuries, the great period of glass expansion in Europe, glass-making practically disappeared in Japan.

The Portuguese and the Dutch brought soda lime and lead glass to Japan. Objects made with such glass were of utilitarian value and had none of the earlier religious aura attached to them. Curiosity, however, led to people attempting to imitate them. By the early nineteenth century fine glass objects were being made. Various feudal lords, in particular Satsuma, were experimenting with glass manufacture. The disruption caused by the arrival of American and other foreign powers who menaced Japan from the middle of the century led to the virtual disappearance of glass manufacture, but private glass shops continued to exist in Tokyo.

One thing which was evident to visitors to Japan in the

seventeenth century was that the superb technical craftsmanship of the Japanese, when applied to glass, produced wonderful objects. The great historian of Japan, Engelbert Kaempfer, who lived in Japan for some years in the later seventeenth century, recorded the expansion of glass manufacture, and that the Japanese were blowing glass by that date. Speaking of Tokyo, he wrote that 'On both sides of the streets are multitudes of well furnish'd shops of merchants and tradesmen, drapers, silk-merchants, druggists, Idol-Sellers, booksellers, glass-blowers, apothecaries and others.' Towards the end of the eighteenth century, Thunberg also noted their ability. 'They are likewise acquainted with the art of making Glass and can manufacture it for any purpose, both coloured and uncoloured.' In the middle of the nineteenth century, during Elgin's mission, there was the same double theme in the accounts. The Japanese could make wonderful glass, but used it only in a restricted way. Despite all their ability, the Japanese used the substance almost entirely for ornamental purposes, harking back in a way to the eighth-century peak of Japanese glass-making. The chronicler of Lord Elgin's mission made the point emphatically:

> It is singular that while the Japanese have brought the manufacture of glass to such perfection in certain forms – as, for instance, the most exquisitely-shaped bottles, so light and fragile that they seem as though they were mere bubbles, of every shade of colour, and beautifully enamelled with devices – plate-glass is unknown among them. Their looking-glasses are circular pieces of steel, polished so highly as to answer all the purposes of a mirror, and usually elaborately ornamented on the back.

Thus two of the major uses of glass in the west, windows and mirrors, were not developed, even by the 1850s.

After the Meiji restoration (1868), there was a flood of new technologies and uses. The multiple uses of glass for windows, lamps and many other purposes were well known. Foreign experts were brought in and industrial glass production soon developed. The result is that Japan is today one of the world's largest producers of glass, probably ranking second after the United States, producing, among other things, a vast amount of plate glass of very high quality.

The puzzle is this. The Japanese had all the knowledge to make high-quality glass from at least the eighth century and probably before. They did make wonderful glass objects, out of coloured and plain glass, but almost entirely for religious or decorative purposes. After about the twelfth century glass-making died out. After the Portuguese reintroduced glass objects they were used for a narrow range of decorative purposes. The Japanese even use the western derived 'girasu' for glass. We have to explain a set of absences.

As we saw in the earlier discussion of mirrors and the concepts of the individual, Japanese mirrors were traditionally made of brass or steel. They were not made of glass. Mirrors were widespread and very important sacred symbols, but glass mirrors were not developed, probably because they were not needed. Thus a whole dimension of perception, the mirror worlds so important in art and science, was more or less absent in Japan. In many contexts, such as in Shinto shrines, the mirror was used not to see the physical body. It was a sacred object through which one could look into the soul. The development of good flat glass for mirrors did not occur before the later nineteenth century.

One of the things that particularly struck Europeans

about Japan was the absence of glass windows. In the later eighteenth century Thunberg noted that 'window-glass which is flat, they could not fabricate formerly. This art they have lately learned from the Europeans ...' Thus there 'are no glass windows here'. He also noticed the absence of mica and mother of pearl. Instead, there were wood and paper screens.

There are several possible reasons for this absence. Firstly, the Japanese climate, on the whole, makes glass windows undesirable. For the very hot and humid half of the year they would have made the tiny house interiors very oppressive. Nowadays, it is only air-conditioning that makes many offices bearable. Secondly, the geology makes glass windows very dangerous, unless made of toughened glass. There are almost daily earth tremors and frequent quakes in many parts of Japan, which would have shattered early glass windows. Then there are the building materials; the flimsy wood and bamboo structures are not suited for glass windows, unlike the brickwork of Europe. There is the question of alternatives. Movable screens made of the superior (mulberry) paper of Japan which lets through light but not wind is an excellent alternative to glass. Another factor was probably cost. Glass is expensive; only a wealthy middle-class family could afford it. Until recently the majority of the population in Japan could not have afforded glass windows.

In the Japanese numbering system, which classifies objects into various categories, one class consists of liquids drunk with a container, as water, wine, tea, etc. The classifier ending is *hai* 'for glasses or cups of any liquid'. In fact, what is very striking is the absence of drinking glasses in Japan. Much of the important development of European glass (in Venice and elsewhere) was in the making of drinking glasses, a continuation of its use in Roman times. Yet in Japan, drinking with glass seems, until the middle of the nineteenth century, to be more or less totally absent. Why? Again there are several obvious reasons.

One concerns the nature of the drink. The Venetian glass was developed for the highest-status and ubiquitous cold drink – wine. In northern countries, where beer was the main drink, it was not drunk from glass, but pewter and pottery. Wine and glass seem to go together. One drinks with the eyes, as well as with the lips, and the glass enhances the effect. Certainly, if one is drinking very large quantities of hot drinks, hot tea, hot water, hot sake, then glass is a bad container. It will crack and the situation is made worse by the fact that thick glass (as was early glass) cracks more easily than thin.

A second, and obviously related fact, is the development of ceramics. With such fine ceramics and wonderful pottery, who needs glass for drinking vessels? Indeed, glass is hardly needed for any other utensils; bottles could be replaced with pots, bowls made with clay. So, apart from the copying of precious stones and a little very fine cut Satsuma ware, glass was not developed for drinking or other utensils. As to the question of the development of glass as an aid to sight, that is lenses, prisms, spectacles and so on, there was no noticeable progress in this direction until the eighteenth century.

The material which we call glass with all its functional associations faded away in Japan, just as it had done in India and China. It was of little use, except for beads, toys and pretty things: aesthetics and ritual yes, practical functions no. Earlier we argued that the unintended consequences of the presence or absence of glass on science, art and personality in the west were probably immense. It does not seem too far-fetched to argue that the well-known fact that at the two ends of Eurasia very different cosmologies and ideologies developed, partly reflected the fact that at one end of the continent a glass civilisation emerged, and at the other a pottery and paper one.

The story of what did not happen outside western Europe has an interesting theoretical implication. For most of history there has been little reason to develop clear glass. Consequently, there is little point in trying to build careful arguments as to why the making of uncoloured glass did not develop. There was no reason why it should develop. It is only when we look backwards at history from our latter-day perspective and see the enormous, but originally subtle, difference that glass has made to the western world that we wonder at its non-existence in other civilisations. History written in a rear-view mirror has its dangers. For, until the very recent advances in reliable knowledge had been made fortuitously as one accidental aspect of the presence of glass, there is not the slightest reason why we should be surprised that glass did not develop much in India, China or Japan.

The major use for glass until the last few hundred years had been for containers. The Chinese and Japanese had excellent containers made of clay. So they had no reason whatsoever to pursue glass. Not only were the consuming public content with the wealth of porcelain and pottery containers, but the producing workers, the vast empire of workshops and potters, were hardly likely to argue that their skills be made less central in order to introduce a technology which requires a lot more fuel (because of the cost of keeping glass molten for long periods) and produces a less robust and arguably less beautiful object. Glass-making is a different technique and there is no great reason why another group should start to do it. What glass there was in most civilisations was mainly used for coloured beads. Clear glass, which would later become the essential kind for use as a tool to see the world in a different way, was for a very long period of little obvious use. So it is, in some senses, a non-question as to why the east Asians did not have glass.

Yet if one says that there were perfectly good containers in the Far East, this still implies that containers in western Eurasia were such that a new substance could be developed alongside it. In the ruthless competition for a niche, there was something about western pottery which meant that it was possible to develop glass-making. The answer might lie in the relative quality of the ceramics, it might be related to the uses to which containers were put (for example, hot and cold drinks), it might be in the organisation and status of the workers in pottery. For example, if the potters had a fairly low status and were then faced with outside glass-workers from the Middle East who had a high status and were well paid, some of them might be pushed aside or change their occupation. On the other hand, the high status of Japanese and Chinese ceramics workers is well known.

To make the story more complex, we have to remember that the factors suggested above are interconnected. It was partly the excellence of their products which gave the potters of the east their high status. That excellence in turn was largely accidental. To a considerable extent Chinese pottery has grown to its predominant position because of the fortuitous presence of two different materials in China. There were large deposits of kaolin and petuntse near to each other.

The kaolin provides the body of the object, the petuntse acts as a flux which will cause overglaze colours to vitrify. It was hence possible to make an excellent hard, dense, beautiful, translucent ceramic. Potters were using the clays that were around them and found that they produced a wonderful substance which we call 'china'. The original discovery of porcelain itself was probably the result of the accidental presence of 'natural' porcelain in China. The resulting ceramics were so desirable that Europeans spent immense fortunes on buying chinaware. The makers of such a fine substance had a high status.

Meanwhile in western Europe these substances were not available, either in the same high quality or quantity. Instead there were other clays out of which a less sophisticated pottery tradition emerged. So it was a matter of luck as to where certain clays were to be found. Thus the different trajectories go back very early, at least to the Roman period. Rome, and through her medieval Europe, opted for pottery and glass, China and Japan for ceramics and paper. Once the divergence had begun it was self-reinforcing. It became more and more difficult to change track. So if one asks why the Chinese did not develop clear glass, one should equally ask why the Romans did not make porcelain. It is only after the event that absences, paths that were not taken, seem so odd. It is thus not surprising, but its effects on different civilisations have been, in the end, immense.

7

The Clash of Civilisations

Who sees with equal eye, as God of all,
A hero perish, or a sparrow fall,
Atoms or systems into ruin hurl'd
And now a bubble burst, and now a world.

Alexander Pope, 'An Essay on Man'

LET US LOOK at what happened when the glass-filled world of western Europe impinged on Asia from the sixteenth century onwards. We can put the case of India on one side. At first, the need for sophisticated glass, for example mirrors and spectacles, was supplied by the hovering European colonial powers. Later, any potential for a large, independent glass-making industry was undermined as India became incorporated into the British Empire. We shall therefore look at the two cases where largely independent civilisations in the east had taken a different path to western Europe and then suddenly found themselves visited by missionaries and traders who brought in new glass goods and the scientific and artistic systems which, we have argued, glass helped to generate. By comparing the Chinese and Japanese cases we can see further complexities in relation to the development and impact of what we tend to assume is a superior technology.

Let us start by looking at the relatively simple matter of the introduction of western glass technology into China from about AD 1600. Although there are signs of earlier western influence, the dramatic change is usually attributed to the emperor K'ang-hsi (1661–1722). He had probably seen examples of western glass brought to his court and in the 1680s or a little later established a specialised glass workshop in the palace workshops, superintended at first by Jesuit missionaries. The techniques used were largely of European origin.

Quite soon the Chinese were making very beautiful and serviceable objects. As the eighteenth-century Jesuit writer Du Halde put it: 'they imitate, well enough, any Pattern that is brought them, tho' they never saw it before. Thus at Present they make Watches, Clocks, Glass ... and several other things which they had no Notion of formerly, or made but very imperfectly.' Recent research suggests that all glass in this period was made in the same area and principally consisted of jars, bowls and cups. The emperor's son moved glass production to the province of Shantung, probably because of the availability there of sand, potash, coal and quartz. When the missionary Alexander Williamson visited that area in 1870 he found that glass was 'regarded as an old established craft in the region, with a number of furnaces in and around the main settlement, supplying dealers in Peking with window glass, bottles of various sizes, moulded cups of every description, lanterns, beads and ornaments, as well as rods of plain and coloured glass sold in bulk, presumably for lampworkers and decorative additions'.

The move of the glass factory away from Peking, and the fact that probably only one limited area of China made glass, explains the account of the decline of glass-making given by

Gillan, who accompanied Macartney's Embassy at the end of the eighteenth century. He observed that much of the glass in China had been imported from the west. This was because, according to him, glass manufacture had ceased in China by the later eighteenth century. 'There was formerly a glass manufactory established at Pekin [sic] under the direction of some of the missionaries, but it is now neglected and no glass is made in China.' Yet glass was used.

> The Canton artists, it is true, collect all the broken fragments of European glass they can find, which they pound and melt again in their furnaces; when melted they blow it into large globes or balloons which they afterwards cut into pieces of various shapes and magnitude as they want it. The chief use they make of it is for small looking glasses and a few toys. This is the only kind of glass they now make in China, and as they blow it extremely thin ...

He ends with the comment that 'they do not seem to understand the manufacture of glass from the crude materials, nor to know exactly what they are. The glass beads, and buttons of various shapes and colours, are imported to them from Europe and chiefly from Venice.' Certainly it seems that glass-making had declined again, though there are some inconsistencies in the accounts which reflect the difficulties foreigners had in understanding this vast empire.

Another indication of the decline in the eighteenth century is the history of paintings on glass. These paintings became an important and valuable export industry in eighteenth-century China, being executed with great skill and artistry on the back of the glass. Again it was probably the Jesuits who introduced the technique and the glass upon which the paintings were made was imported from the west.

What is particularly fascinating and relevant to the question of the different development of reliable knowledge at the two ends of Eurasia is the impact of European glass instruments of measurement and sight. In 1738 Du Halde published a number of Jesuit accounts from original letters and books of the seventeenth century. The curiosity of the Emperor K'ang-hsi led the missionaries to show off their wares. 'They first gave him an Insight into Optics, by presenting him with a pretty large Semi-Cylinder of a very light kind of Wood; in the middle of whose Axis was placed a Convex-Glass, which being turned towards any Object exhibited the Image within the tube in its natural Figure.'

This was but one of their displays of what had been discovered in the west. Father Grimaldi

> gave another Instance of the Wonders of Optics in the Jesuits Garden at Peking, which greatly astonished all the Grandees of the Empire. He made upon each of the four Walls, a Human Figure of the same Length as the Wall, which was fifty Feet: As he had strictly observed the Rules, there was nothing seen on the Front, but Mountains, Forests, Chaises, and other things of this Nature; but from a certain Point you perceived the Figure of a Man, handsomely shaped, and well proportioned.

The role of glass, and particularly mirrors, was especially important. 'It would be too tedious to mention all the Figures that were drawn confusedly, and yet appeared distinctly from a certain Point, or were reduced to order by help of Conic, Cylindric, and Pyramidical Mirrors; together with the many Wonders in Optics, that P. Grimaldi exhibited to the finest

Genius's in China, and which equally excited their Surprise and Admiration.'

Other instruments of glass were also displayed.

> In Catoptrics they presented the Emperor with all sorts of Telescopes and Glasses, for making Observations of the heavens and on the Earth, for taking great and small Distances, for diminishing, magnifying, multiplying, and uniting Objects. Among the rest, they presented him first with a Tube made like an octagonal Prism, which being placed parallel with the Horizon exhibited eight different Scenes, and in so lively a Manner that they might be mistaken for the Objects themselves; this, joined to the Variety of Painting, entertained the Emperor for a long time. They next presented another Tube, wherein was a Polygon-Glass, which by its different Faces collected several Parts of different Objects to form an Image; so that instead of Landships, Woods, Flocks, and a hundred other things represented in the Picture, there appeared a human Face, an intire Man, or some other Figure in a very distinct and exact manner.

Nor was it just instruments of vision which were displayed. 'They likewise presented the Emperor with Thermometers, to shew the several Degrees of Heat and Cold. To which was added a very nice Hygrometer to discover the several Degrees of Moisture and Dryness ...' The conclusion of all this was very gratifying for the missionaries for the Chinese were checked in their sense of superiority. 'All these different Inventions of Human Wit, till then unknown to the Chinese, abating somewhat of their natural Pride ...'

What is more difficult to prove, though at a very general level it seems so obvious, is the way in which this contrast in the availability of scientific instruments made with glass affected knowledge at the two ends of Eurasia, both in what we call science and in art. Du Halde gives us some clues. The Jesuits used some of the glass instruments to show how their knowledge was greater. They did this not only in the ways described above but also at some length in proving that their astronomy was superior. Du Halde implies that mathematics was far less developed than in the west, perhaps somehow related to the Chinese's backwardness in optics. 'As for their Geometry, it is superficial enough; for they are very little versed, either in the Theory, which demonstrates the Truth of Propositions called Theorems, or in the Practice, which teaches the method of applying them to Use by the Solution of Problems.' The 'other Parts of Mathematics, excepting Astronomy, were entirely unknown to the Chinese; nor is it above a Century since they began to perceive their Ignorance upon the Missionaries' first Arrival in China'. He also implies that the mathematics embedded in the visual arts was related to glass – the sense of perspective. All this is fascinating, if overdrawn.

Immense as the implications are of what he wrote, these aspects of glass are only part of the picture. Du Halde hardly alludes to many other important developments. The implication of microscopes is overlooked. The effects of glass on chemistry are not mentioned. Nor is he able to address the question of what had happened to reliable knowledge in China in the preceding centuries. There is the curious fact that while the Chinese made considerable advances in optics very early on, reputedly reaching by the thirteenth century a similar level of sophistication and knowledge as the Greeks, they halted there. They did not then make the breakthrough which was achieved first in Islamic optics and was then built on in western Europe. So recent authorities conclude that ancient Chinese optics was based on empirical observations and short on theoretical abstraction or quantitative description. Consequently when the fruits of the Arabic and western optics, based on experiments with glass, were introduced to China from the seventeenth century the foundation of traditional Chinese optics was altered.

The impact of western glass technologies is given another twist if we look at the one further case of Japan. It shows three things. Firstly, it confirms the absence of scientific glass in China up to the arrival of the Jesuits. Japan was largely dependent on China for its technology and usually quick to emulate it. If there had been glass instruments on a wide scale in China, they would have been imported. Yet glass was brought in by westerners, in particular the Portuguese and the Dutch. This is further evidence as to the non-development in China.

A second inference is the way in which it takes more than artefacts to change a knowledge culture. In China the impact of the Jesuits with their marvels was almost nil; the clocks and glass tools were mainly kept as curious museum pieces and had hardly any impact for several hundred years. Yet the Japanese for centuries had been keen to import new ideas and technologies from their giant neighbour and this may help to explain why they were so fascinated by western glass instruments and very rapidly absorbed them thereby altering their knowledge of the world.

Finally, Timon Screech has recently minutely examined the case of Japan, specifically in relation to scientific instruments of glass. His well-documented case gives us the opportunity to see some of the deeper ways in which a gradual and hence largely invisible process of knowledge accumulation, which had occurred over hundreds of years in the west, made a difference. The encounter was sharp and relatively short and some of its salient characteristics therefore stand out. The encounter of visual systems also reminds us that the non-development of glass has nothing to do with technical ability. As soon as they saw a use for it, as had earlier happened with guns, the Japanese made excellent glass.

The technical ability was such that when necessary the Japanese could apply glass to any purpose. For instance, the making of scientific instruments, when required, was no problem. In the 1790s Thunberg observed that 'In like manner they understand the art of glass-grinding and to form Telescopes with it, for which purpose they purchase mirror-glass of the Dutch.' As for microscopes, Screech describes how these very soon became a symbol of western scholarship (*Rangaku*). They were imported from the seventeenth century onwards and were known as *mikorosukopyumu*. One account by a Japanese who looked through one captures the sense of amazement.

> We brought several things into focus and inspected them under it. The clarity of the minutiae was quite extraordinary. Salt crystals could be seen to have a hexagonal shape, while buckwheat flour (even the most finely sifted sort) was triangular. A candle wick looked like a loofah and mould was shaped like mushrooms; water was like hemp leaves

> with patterning on them, ice had a warp-and-woof design; sake was like boiling water, all seething in bubbles . . .

As in the west, an invisible world beneath the surface of reality suddenly emerged, framed and focused by the microscope. The early microscopes were often displayed at roadside fairs, and manuals for their use were published in Japanese.

Glass was increasingly used in chemistry for retorts, dishes, flasks and tubes. It was also used to store specimens. We are told that the technical-looking vessels were called *forsake* (flasks), whereas goblet shapes were *koppu* (cups). The Japanese, unlike many others, quickly moved on from the wonder and fascination of glass to its utility. As an eighteenth-century Japanese writer observed, 'Initially, the material was enjoyed just for its sparkle and shine, but of late it has been recognized that glass ought not to be limited to use in playthings. Jars and bottles have been made, and things stored inside them. When so kept, a material's original characteristics (*honsei*) are preserved indefinitely; medical substances or fragrances can be passed on like this over long periods.' Another use for glass was for spectacles, which were increasingly worn, though they clashed with Japanese etiquette since they led to rather direct and rude staring at others.

It would clearly be foolish to argue that all precise scientific or artistic knowledge is dependent on glass. The magnificent increases in reliable information in early China up to the fourteenth century have been well documented by Joseph Needham and others. The magnetic compass, the quadrant, the astrolabe, even the mechanical clock, do not need glass. But it seems equally true that without this mysterious substance many avenues are blocked. The enthusiastic reaction of the Japanese and the lukewarm one of the Chinese illustrate well how much its use is determined by non-obvious cultural and social factors. What is not in doubt is that those who took part in the great confrontation of civilisations between the sixteenth and nineteenth centuries and who wrote about these topics bore testimony to the fact that the increasingly glass-dominated world of the west had encountered civilisations which had effectively given up the use of this material. The astonishment of the Chinese mandarins as they assembled round the Jesuit scientific instruments and perspective drawings is an epitome of how far the two ends of Eurasia had drifted apart.

The process of drifting apart of the two ends of Eurasia and their subsequent forceful meeting can also be seen in the history of art in China and Japan. An account of *The Meeting of Eastern and Western Art* by Michael Sullivan reveals some of the differences between post-Renaissance western art and east Asian art. That the revolutionary change in perspective and realism occurred only in the area where glass artefacts were common and not in China and Japan is, we argue, more than just a coincidence.

Sullivan quotes various early accounts by western visitors. An Arab merchant who visited China in the ninth century reported that 'the Chinese may be counted among those of God's creatures to whom He hath granted, in the highest degree, skill of hand in drawing and in the arts of manufacture'. Marco Polo in the thirteenth century also noted a castle whose hall was decorated 'with admirably painted portraits of all the kings who ruled over this province in former times'. Fifty years later Ibn Battuta wrote that the 'people of China of all mankind have the greatest skill and taste in the arts. This is a fact generally admitted: it has been remarked in books by many authors and has been much dwelt upon. As regards painting, indeed, no nation, whether of Christians or others, can come up to the Chinese, and their talent for art is something extraordinary.' Sullivan notes, however, that Battuta was not describing the paintings of the literati, but the works of professional painters at the frontier stations who were employed to make likenesses of visitors to the country.

In earlier times, this skill included knowledge of some of the key techniques used in Renaissance art. Early Buddhist art in central Asia had included ways of modelling in two tones and the use of shading to suggest relief. This had been imitated to a certain extent in both China and Japan. But the methods were always looked on as a foreign technique and as Buddhism's power faded, so did the elements of chiaroscuro and perspective which had existed on the edges of east Asian art. A particularly dramatic example of this fading away of skills which were not so far distant from the developments in Renaissance Europe is a famous example in early twelfth-century China.

Foreshortening, shading and the receding ground line appear in a very realistic handscroll called 'Going up river at Ch'ing-ming festival time', painted by Chang Tse-tuan between AD 1100 and 1130. There is three-dimensional space, but the painting represents the way in which a moving eye would see the scene, rather than using the fixed single-point perspective which developed in the west. Yet even this move

10. Pines and rocky peaks

These two pictures, painted some hundred years after the 'Going up the river...' picture on page 52, show far less realism and more imagination and symbolic representation, following later stylistic conventions. 'Going up the river' was made at the scene, these two are from memory. The earlier painting shows both foreground and background in equal detail, these two concentrate on the foreground and leave the background very hazy, a widespread characteristic of Chinese painting of later periods. The early painting could only have been done by someone with good distant vision; these two could have been painted by someone with short-sight (myopia) who had seen similar paintings from close up.

towards perspective was abandoned. Likewise, there are examples of oil paintings, such as the decoration of the Tamamushi Shrine in the eighth century. But this was also abandoned. Realism and mirror-like representations were considered vulgar and not suitable for a scholar-painter. The Chinese artist, Gong Xian (1619–89), quoted by Clunas as follows, explained the distinction:

> In ancient times there were pictures (*tu*) but no paintings (*hua*). Pictures depict objects, portray people, or transcribe events. As for paintings, the same isn't necessarily true for them ... [To do a painting], one uses a good brush and antique ink, and executes it on a piece of old paper. As for the things [in a painting], they are cloudy hills and misty groves, precipitous boulders and cold waterfalls, plank bridges and rustic houses. There may be figures [in the painting] or no figures. To insist on a specific subject or the representation of some event is very low class.

Thus the aim of painting was not a realistic representation of the natural world, but to convey something of the deeper, spiritual essence. The idea was not to imitate nature but to use art as a way to communicate through symbols to the heart and feelings of the viewer.

Both Chinese and westerners were thus surprised when the famous missionary Matteo Ricci brought Renaissance artworks to China at the start of the seventeenth century. Sullivan quotes Ricci to the effect that 'Chinese use pictures extensively, even in the crafts, but in the production of these and especially in the making of statuary and cast images they have not at all acquired the skill of Europeans. They know nothing of the art of painting in oil or of the use of perspective in their pictures, with the result that their productions are lacking any vitality.' His views, self-vaunting though they are, are further reported by Gu Qiyuan, writing in a book published in 1618, after Ricci's death, and quoted by Clunas. He described a painting brought by Ricci. 'This Lord of Heaven is painted as a small boy, held by a woman called "Heavenly Mother". He is painted on a copper panel, with five colours spread on top. The face is as if living, the body, arms and hands seem to protrude from the panel, the concave and convex parts of the face are no different from those of a living person to look at.' When asked how the painting could achieve this, Ricci replied,

> Chinese painting only paints the light (*yang*), it does not paint the shadow (*yin*). Thus to look at, people's faces are completely flat, with no concave or convex physiognomy. My country's painting combines the *yin* and *yang* in drawing, so that faces have higher and lower parts, and arms are round ... The portrait painters of my country understand this principle, and by using it are able to ensure that the painted effigy is no different from the living person.

As Chiang Shao-shu, quoted by Sullivan, commented a century later on the same painting, it is of 'a woman bearing a child in her arms. The eyebrows and the eyes, the folds of the garments, are as clear as if they were reflected in a mirror, and they seem to move freely. It is of a majesty and elegance which the Chinese painters cannot match.'

Yet the Chinese literati were not persuaded by Ricci to change their ancient traditions and seem to have forgotten the whole episode. Only this can explain the sense of surprise when the Jesuits displayed their art to the emperor some seventy years later, as described by Du Halde:

> Nor was Perspective forgotten: P. Bruglio gave the Emperor three Draughts performed exactly according to rules, and he hung up to View three Copies of them in the Jesuits' Garden at Peking. The mandarins, who flocked to this City from all Parts of the Empire, came to see them out of Curiosity, and were all equally surpriz'd at the Sight; they could not conceive how it was possible on a plain Cloth to represent Halls, Galleries, Porticos, Roads, and Avenues reaching as far as the Eye could see, and all this so naturally as at the first View to deceive the Spectator.

The difference between artistic traditions was clearly immense and well recognised by the Chinese experts. For example, in the early eighteenth century Sullivan quotes the landscape painter Wu Li on some of the differences between Chinese and western art. 'Our painting does not seek physical likeness [*hsing-ssu*], and does not depend on fixed patterns; we call it "divine" and "untrammelled". Theirs concentrates entirely on the problems of dark and light, front and back, and the fixed patterns of physical likeness.'

Once again, though, the effects of another wave of western artistic missionising were negligible. The techniques of realism, perspective and shading moved down to the lower levels of the craftsmen painters, but were of little interest to the literati. From the later seventeenth century until the middle of the nineteenth century they ignored the western tradition. The court artist Tsou-I-kuei, quoted by Sullivan, summarised the fatal flaw in western art:

> The Westerners are skilled in geometry, and consequently there is not the slightest mistake in their way of rendering light and shade [*yang-yin*] and distance (near and far). In their paintings all the figures, buildings, and trees cast

> shadows, and their brush and colours are entirely different from those of Chinese painters. Their views (scenery) stretch out from broad (in the foreground) to narrow (in the background) and are defined (mathematically measured). When they paint houses on a wall people are tempted to walk into them. Students of painting may well take over one or two points from them to make their own paintings more attractive to the eye. But these painters have no brush-manner whatsoever; although they possess skill, they are simply artisans [*chiang*] and cannot consequently be classified as painters.

As Sullivan comments, 'to the Chinese gentleman-painter who aimed at a triple synthesis of painting, poetry and calligraphy, what had the laborious realism of oil painting to do with fine art?'

This preservation of an ancient tradition, which in many ways was similar to the early medieval art of the west, gives us some indication of the force that was needed in the west in the period between Giotto and Leonardo in order to change the whole artistic tradition. It does not seem implausible to argue that without those glass-based developments which were central to western optics there would likewise have been no revolution in the west.

Japanese art is just as ancient a tradition as that of China, stretching back for more than fifteen centuries. Like Chinese art, classical Japanese art was not dedicated to realism, but rather to conveying deeper truths through symbolism. Many of its central features were well described by Henry Bowie. Artists did not aim at 'photographic accuracy or distracting

11. Two views of Derwentwater

Made from a similar vantage point, these two representations of the lake and mountains show quite different treatments of the distant view. The Chinese artist, working in the Chinese tradition, produces a quite indistinct, stylised, and blurred background, such as would be 'visualised' by many Chinese viewers.

detail', and they painted what they felt rather than what they saw. While the artist was often encouraged to study minute details, including insects, flowers, birds, fish and so on, these objects were then incorporated through a set of standardised techniques which were not meant to reproduce an actual natural scene.

There were numerous rules. Low mountains in a landscape suggested great distance, thus Mount Fuji should not be painted too high or it would lose in dignity by appearing too near. There were eight different ways in which rocks, ledges and similar features should be represented. Garments, their folds and lines could be painted in eighteen different ways. It is clear that the artist was very constrained by a code of how things should be represented. The natural world provided some clues, but to a large extent the art consisted in applying various theories and codes.

A further constraint on the paintings was the fact that the medium was water-based. A soft brush was used on very absorbent paper, so that paintings had to be executed very fast, and in fact were scarcely distinguishable from writing or calligraphy. No easels were used and the artist would sit on the floor with the paper or silk spread before him on soft material. The technique in the hands of a master is described in a quotation from Bowie as follows. 'In landscape work the general rule is to paint what is nearest first and what is farthest last. Kubota's method was to do all this rapidly and, if possible, with one dip of the well-watered brush into the *sumi*, so that as the *sumi* becomes gradually diluted and exhausted the proper effect of foreground, middle distance and remote perspective is obtained.' This was entirely different from the oil paintings of the west, where mistakes could be corrected and details added.

As well as having to be done rapidly and without hesitation, the painting was also done from memory, not from reality. Otto Rasmussen, who lived for many years in China, described how Chinese landscape painters did not paint from life. 'They simply wandered about or sat in meditation and then went back to their studios without a sketch or any sort of rough drawing to paint their pictures.' This perhaps helps to give part of the context for a story told by Gombrich in *Art and Illusion*: 'James Cheng, who taught painting to a group of Chinese trained in different conventions, once told me of a sketching expedition he made with his students to a famous beauty spot, one of Peking's old city gates. The task baffled them. In the end, one of the students asked to be given at least a picture postcard of the building so that they would have something to copy.'

Into this ancient tradition came the western visitors with their geometrical perspective and realistic art, their chiaroscuro effects and attention to accurate detail and proportion. The reaction of the Japanese was, however, not identical to that of the Chinese. While they recognised the huge gulf between their high art and the western Renaissance tradition, they were more attracted to the western techniques than were the Chinese and were successful in adopting them on occasions.

Sullivan provides a number of examples of successful perspective paintings by Japanese artists. In the late eighteenth century Kokan was fascinated by western realism and even made a *camera obscura* to help in creating perspective drawings. He wrote that 'If one follows only the Chinese orthodox methods of painting, one's picture will not resemble Fuji.' There was only one way out. 'The way to depict Mount Fuji accurately,' he declared, 'is by means of Dutch painting.' During the eighteenth century perspective pictures became quite widespread in Japan. There are notable examples in some of the works of two of the most widely known Japanese artists, Utamoro and Hokusai.

We could end the story here, for a number of reasons have been given to explain a different trajectory at the two ends of Eurasia. The aim of real painting in China and Japan was to create a surface which conveyed symbolic and deeper meanings, rather than achieve a photograph-like mirror of the outward forms of nature. The materials used, absorbent paper laid flat and large ink-filled brushes, encouraged a swift and flowing execution, based on memory and rules. These conventional rules as to how effects were to be achieved were known to artist and viewer alike and dictated what should be in the picture and how it should be portrayed. All this gives us, it would seem, a sufficient explanation. One sees the rules, one sees the paintings. Similar purposes, techniques and rules had, after all, covered the entire world, including western Europe, up to the fourteenth century. They account perfectly well for the artistic productions that have survived. There is no need to suggest that the Greeks or the medieval painters of icons painted in that way because of any particular feature of their eyes.

Yet in the Chinese case there may be an almost invisible second factor. It has been invisible because there was hardly any need to invoke it. All was explained, it seemed. Yet it may be a reinforcing or secondary reason lying alongside the other factors. Rasmussen suggested that 'The theory that Chinese painters, who were nearly always literati poets and philosophers ... painted the spirit rather than the material, would not altogether account for the invariably minute, almost photographic foreground detail, and the misty background of their pictures.' He notes that Chinese artists 'pictured near objects in detail and their distance as a more or less complete blur'. If we try to explain this by alluding to some idea of 'spiritual impressionism' it 'does not explain why they were able or wanted to "memorise" and depict so many of the nearest objects in such material detail. Nor does it explain why their spiritualistic conceptions are confined almost invariably to background and not to foreground subjects.'

What he intriguingly suggests is that these same literati were, in effect, making a virtue of necessity. Many of them were very short-sighted and this meant that before spectacles were invented they could not have painted in the way in which Van Eyck or Leonardo did, even if they had wanted to do so.

Before we dismiss this as a totally absurd suggestion it is worth considering further. If one is myopic, or if not, if one puts on myopia-simulating glasses, one can see that the only way in which one could paint if one is myopic is the way in which the Chinese did in fact paint. One has to rely almost entirely on one's memory and inner eye, through a formula, through making a virtue of standardised representations (preferably hazy) of larger objects. The only way is to go up very close to individual objects (including other paintings) and then come back to one's workplace and carry out the painting from memory. What would be painted would have to be an idealised montage with the details of insects, flowers or people. Furthermore, if one has a large scroll and one can only see about ten inches clearly, the further part of what one is painting is already blurred. Finally, when the painting was unrolled (rather than being hung on a wall as in the west), if the high-class viewer was also short-sighted he would see in the painting what he had experienced in life – detail in the foreground, a blurred background. Indeed, he might not really see the picture at all well. This gives an added twist to the complaint of an old artist, quoted by Binyon, that 'People look at pictures with their ears rather than with the eyes.'

Chinese and Japanese art is famous for the strange hazy portrayal of middle and far-distance objects. In this respect it is quite different from the art of the other great traditions such as Islamic or Byzantine painting. The congruence of this mysterious and underdefined representative style with what one would find in elite artists who were often myopic is intriguing. The degree to which it is related obviously rests on the degree to which there is evidence of unusual levels of short-sightedness in China and Japan. If such evidence exists, it would add a final twist to the story of diverging vision, for it would reinforce the fact that western eyesight was made more powerful by glass by suggesting that eastern eyesight was facing a large-scale challenge in seeing objects at a distance. We will notice once again the comparative difference between civilisations where glass was widely available and those where it was little used.

8

Spectacles and Predicaments

Why has not Man a microscopic eye?
For this plain reason, Man is not a Fly.
Say what the use, were finer optics giv'n,
T' inspect a mite, not comprehend the heav'n?'

Alexander Pope, 'An Essay on Man'

IT IS ONE of the ironies of life that just as they reach the peak of knowledge, in their late forties and fifties, many people find it impossible to continue reading without glasses. They have to hold what is to be read at such a distance away from their eyes that they cannot distinguish the characters. This was a serious drawback up to the fifteenth century, especially for bureaucracies and companies where the most skilled in literacy and accounting had to give up work as their eyes failed. It became an even more serious disability after the printing revolution made books for scholarship or private enjoyment widely available. It may therefore strike many as no surprise that it is exactly at the period of growing wealth and bureaucracy that the making of spectacles developed rapidly. The eyeglasses made of two bi-convex lenses suspended on the nose to help those with old age long-sight (presbyopia) were probably invented at around AD 1285 in northern Italy and their use spread rapidly in the next century, so that spectacles were a widespread feature of European life half a century before movable metal printing was invented by Gutenberg in the mid fifteenth century.

The effects of this development in western Europe were immense. The invention of spectacles increased the intellectual life of professional workers by fifteen years or more. As a number of historians have pointed out, the revival of learning from the fourteenth century onwards may well be connected to this. Much of the later work of great writers such as Petrarch would not have been completed without spectacles. The active life of skilled craftsmen, often engaged in very detailed close work, was also almost doubled. The effect was both multiplied, and in turn made more rapid, by another technological revolution to which it was connected, namely movable type printing, from the middle of the fifteenth century. Obviously, the need to read standard-sized print from metal types in older age was another pressure for the rapid development and spread of spectacles, and the presence of spectacles encouraged printers to believe they had a larger public.

The invention and spread of spectacles with glass lenses may be seen as a necessary, even inevitable, development until we step outside western Europe and look to the east. If we do so we are faced with a real puzzle. For, as far as we know, up to about the seventeenth century at the earliest glass spectacles were not developed to any great extent outside Europe. In Islamic civilisations, in India, in China and Japan, the western type of spectacles was practically unknown until Europeans introduced them from the seventeenth century. Why was the development of spectacles confined to one tip of Eurasia for about four centuries? There are several theories. One is that the Chinese, at least, had an alternative natural substance which could be, and was occasionally, used, namely rock crystal quartz. Yet even the rare quartz spectacles that were made seem not to have been lenses; they were just flat slabs used to protect the eyes.

In pursuing this puzzling absence, we can focus for a moment on the Japanese case. The Japanese knew of the concept of putting two pieces of quartz on wires in front of the eyes from their Chinese neighbours. They had made very fine blown glass objects as early as the eighth and ninth centuries. Many Japanese nowadays wear glasses or contact lenses, perhaps one of the highest rates in the world, and it seems unlikely that this reflects a recent change. When spectacles could be manufactured widely, they became very popular. The traveller Isabella Bird throws an interesting sidelight on this in the 1880s: 'The entire police force of Japan numbers 23,000 educated men in the prime of life, and if 30 per cent of them do wear spectacles, it does not detract from their usefulness.' Thus there was a huge demand for spectacles in the later nineteenth century. Furthermore, we know that eye diseases and attention to the eyes were very widespread in Japan, with constant cleaning of the eyes. Yet, as far as we can see, spectacles were practically unused in Japan until the nineteenth century. We are left with the intriguing puzzle of why they were not developed.

The puzzle takes us to China, from where Japan received most of its major technologies before the nineteenth century. The oldest certain accounts of double-lens spectacles using glass are from Ming accounts (between the middle of the fifteenth and sixteenth centuries) and refer to western imports. Earlier references to spectacles were to dark substances (often 'tea' crystal) used to protect the eyes against glare and dirt, for healing (quartz had magical properties) or to disguise the reactions of judges from litigants who appeared before them. It was from the middle of the seventeenth century that spectacles made of glass became fairly widespread.

The tradition that glasses were as important for status and eye protection as to counter the effects of ageing continued up to the end of the eighteenth century. This is shown by Gillan's account when he accompanied the Macartney Embassy of 1793–4: 'The Chinese make great use of spectacles ... The eye glasses are all made of rock crystal.' He continues that 'I examined a great number of polished eye glasses after they were ready for setting, but I could not observe any diversity of form among them; they all appeared to me quite flat with parallel sides. The workmen did not seem to understand any optical principles for forming them in different manners so as to accommodate them to the various kinds of imperfect vision.'

In 1868 the missionary Williamson observed the making of rock crystals into spectacles in Shantung province. Citing his work Hommel states at the start of the twentieth century that 'I was told in Tsingtao that the famous optical works of Zeiss in Jena has procured from the same locality rock crystal for optical instruments.' And he adds that 'Only in very recent times have Chinese opticians revolutionized their trade by the introduction of foreign methods for testing the eyesight and fitting glass lenses according to their tests.' He wonders why, if it is true that European glass spectacles were introduced to China in the fifteenth century, 'the Chinese took the revolutionary step to abandon the use of glass for lenses, and employ instead a material but poorly fitted for the purpose'. So there is another mystery. The Chinese had the idea of a material for protection of the eyes. They also, according to Joseph Needham, had the notion of magnifying lenses. Why did they wait to import the idea of magnifying spectacles from the west and why did they make so little use of spectacles once they had them?

Otto Rasmussen was brought up in China in the late nineteenth century and was moved by the sight of blind beggars in Shanghai. The fierce light of the Chinese sands caused him to suffer from sun-blindness for a while. He trained as an ophthalmic surgeon in America and on the basis of twenty-five years of research in China from 1908 he built up an unrivalled picture of Chinese eyesight.

He believed that the absence of the development of lenses in spectacles to overcome sight defects was partly because of the non-development of glass in China. It would have been strange if they had suddenly leapt from a situation where there was very little use of glass into the kind of experimentation with lenses which we see in the west. This explanation may be sufficient. But we would like to explore a supplementary and rather extraordinary hypothesis which is also suggested by Rasmussen's work, although he himself does not make the link.

This involves looking more closely at the nature of what the eye was expected to read, and variations in eyesight. In relation to the objects of vision, it is worth remembering the difference between printing at the two ends of Eurasia.

Chinese and Japanese printing used wood blocks. These were easily made in different sizes and redrawn as needed. In the west, on the other hand, expensive metal type printing developed leading to the printing of numerous, often fairly rough and small, standard-sized books which would have been very difficult to read in older age without spectacles.

All of this, of course, makes the large assumption that the nature of the eye problem at the two ends of Eurasia was the same, namely presbyopia or the growing inability to see close objects clearly after the mid-forties. This assumption needs to be tested. We know that this was the major problem in western Europe, as it is today. But supposing this was not the great difficulty in eastern Asia? Suppose that the major problem was the opposite, namely myopia (deteriorating far sight). If that were the case, it would have a strong effect on the development of spectacles.

Myopia usually starts to manifest itself in childhood, between the ages of four and ten. To cure this would require making spectacles for a relatively powerless group (children). Furthermore, it would not seem so necessary since in relation to reading or other close work, the difficulty can be overcome by holding things up very close to the eyes. Added to this, the concave lenses which are needed to correct myopia are far more difficult to grind than convex ones. Spectacles for myopia were invented in the west only after some two hundred years of making convex lenses for presbyopia. Finally, as the child grew up, the retina stretched so the myopia might have partially cured itself and sight would have normalised. Thus spectacles are very unlikely to develop in a civilisation whose main problem is myopia.

With people who mainly suffer from presbyopia, the

12. Concave lens for myopia

The earliest depicted concave glasses for myopia. Painting by Jan van Eyck, 1436. Concave lenses were first made in Europe some century and a half after convex lenses for long-sightedness.

opposite is true. Presbyopia characteristically develops from about the age of forty as people reach the peak of their careers. People of such an age are likeliest to have the political power to be needed for bureaucratic and other purposes. They are also likely to have the money to invest in spectacles. The convex lenses they need are much easier to grind than concave lenses which need to be hollowed out. Furthermore, for the long-sighted person or presbyope there is no way round the problem. The object to be seen is either close up and blurred, or far away and unreadable. So did the Japanese and Chinese suffer from unusually high rates of myopia?

What Rasmussen encountered in his many years of working with Chinese eye problems astonished him. He summarised the research of a quarter of a century by saying that 'only 20 per cent of all lenses supplied by the ancients, and similar amount by the moderns (excluding high astigmatic) were for old-sight lenses. It is the other 80 per cent that matters; the 65 per cent for myopia and the 15 per cent for glare and therapeutics.'

The significance of the degree of myopia comes out only if we compare the Chinese results with those for other nationalities. In 1930 a Chinese expert published figures for 569 Chinese and 568 white residents of Peking who had been examined in the previous two years. Some 70 per cent of the Chinese investigated were myopic compared to 30 per cent of the foreigners. Furthermore the degree of Chinese myopia was far higher than that of foreigners. Although unfortunately Rasmussen does not specify the sources and dates of the 'Old Chinese Records', he was convinced that 'myopia is, and has been for centuries', the greatest cause of eye problems. He believed that 'almost a quarter of the nation was, and probably still is, myopic. Seventy-five per cent of all spectacles worn are for short-sightedness.'

Rasmussen's early figures for high myopia in China and the implications for Japan can be checked against more recent data. In 1980 the eye surgeon Patrick Trevor-Roper suggested that while in western countries myopia rates were about 15–20 per cent of the total population, in China and Japan the rates were about four times as high at 60–70 per cent. A recent graph of the incidence of myopia in England shows that by the age of seventeen it is about 15 per cent, and the highest peak is at about forty when it is a little under 30 per cent. This contrasts with rates in the Far East, which are far higher. Specialists refer to rates of 85 per cent of schoolchildren in Taiwan being short-sighted. One of the highest rates ever recorded is in Singapore, where 98 per cent of medical graduates were found to be myopic. In an interview conducted with Dr Takashi Tokoro, an eye specialist in Tokyo, in April 1999, it emerged that, among children aged about eleven, 30 per cent suffer from serious myopia, by the age of fifteen, this rises to 50 per cent, and by university entrance at seventeen, to 70 per cent. Given the fact that there is a high level of 'late onset myopia' as well, it looks as if 80 per cent of the adult population might be seriously myopic.

What stands out very strongly is that certainly in Japan, Taiwan and Singapore the rates of people with myopia are very high, and one suspects the same is true in China. Thus the current situation suggests that myopia may well have been, as Rasmussen argued, the central eye problem in much of eastern Asia in the twentieth century.

What is more problematic is to know how far back this pattern reaches. Apart from the problems of obtaining any hard evidence, there is the added problem of seeing whether the rates were changing. It appears that they were high by the 1920s in China. Had there been a long-term change, and when had it begun? We shall have to proceed by two indirect methods, neither of them entirely satisfactory, but at least providing circumstantial evidence. We have to establish what the probable reasons for the high levels of myopia in eastern Asia are. If these can be discovered we can then check whether they were present in the earlier period. If so, it seems reasonable to argue that myopia may well also have been widespread. The second involves a detour. There has been considerable speculation on the possible effects of widespread myopia, both at the individual ('myopic personality') and wider institutional and cultural level. We can look for some of these effects. This second venture is also important for it links with our wider attempt to see how changes in human vision, whether by glass or in other ways, have shaped cultures, and in turn been shaped by culture.

One of the major theories to account for myopia is genetic. There is clearly some relationship between heredity and myopia. It is known that certain families tend to be short-sighted and others long-sighted, which is not surprising since shape and size of the eye are inherited. Unfortunately, the limitation is that heredity, as usual, is mixed up with a family lifestyle. The complexity is shown by the fact that surveys indicate that identical twins are not usually both myopic.

Particularly interesting as a test of the genetic theory is a well-known study of Canadian Inuit which shows that genetics is not at the root of the matter. Their eyes were examined over three generations. Only 5 per cent of the older generation displayed signs of myopia, whereas 65 per cent of their grandchildren had elongated eyeballs. The research also showed an increase from 2 per cent myopia in one generation to 45 per cent in the next. This occurred without any obvious change in diet or lifestyle. The one change was in education. This suggests that eyestrain through reading is a more relevant factor.

Another theory for the prevalence of myopia is that weakness in the eye caused by malnutrition may be a major background factor. Early in his booklet Rasmussen wrote that blindness and eye troubles in China were due 'above all' to 'malnutrition, due to the absence of Vitamin A in their foods'. He refined his study by looking at the regional incidence of myopia and found that, while myopia was high in all regions, it was 'highest of all in the Yangtze valley and Central China regions'. He wondered whether in this area 'the primitive farming methods, particularly of rice and vegetables, exhaust the Yangtze valley soil more than elsewhere and produce inferior crops'. He suggests that the vegetables in this region are 'neither as tasty nor satisfying as the same things in the West or imported to China'.

At the end of his booklet he comes back to the same theme, drawing attention to the subdividing of plots of land as population built up: 'As this situation developed, the land was over-cultivated and underfertilised.' Rasmussen summarises what he now guesses is something going back over a thousand years: 'In such circumstances, it could not be expected otherwise than that the quality of grain, tubers, and general produce for domestic consumption deteriorated in chemical content and sustenance value.' He then refers to 'the evidence of modern Western and Chinese medical and farming experts that foods were deficient in vitamins and were the direct cause of diseases that sent millions blind or reduced their vision'. If this was true in the last twenty years, he asks how far back did it go: 'I suggest that in the older and more thickly populated areas of the river basins it must have been a developing problem at least 1,500 years ago.'

Several comments can be made about this suggestion. Firstly, it is generally agreed that vitamin A deficiency (as well as other deficiencies such as calcium), does indeed have a serious effect on the eyes. There are numerous studies to this effect. Vitamin A is to be found in liver, eggs, butter, milk and cheese and in fatty fishes and in some yellow vegetables. The traditional mainly vegetarian diet had hardly any of these and even the fish component was usually small. Vitamin A can also easily be destroyed by cooking, and it has been suggested that the frying characteristic of China may affect this. Since we know very little about traditional diets in China we have to be cautious. But if it is true that vitamin A mainly comes from animal sources, it is not difficult to see how the reduction in animal products as the Chinese increasingly relied on rice and vegetables probably had a considerable effect.

The problem, however, is that we are dealing with an effect (myopia) which has a set of interacting causes. The effects of nutrition undoubtedly interact with genetic factors and eyestrain through close work. This can be seen in the present situation in Japan. Currently there are extremely high rates of myopia among Japanese schoolchildren. Yet schoolchildren have free school milk and nowadays eat a lot of meat, cheese and vegetables. This, one might surmise, would rule out vitamin A deficiency. This is clearly the case now, but it may well be that in the past it was an important contributory factor, weakening the eyes, causing difficulty in reading in dark classrooms and hence exacerbating the effects of close work. The fact that the diet is now much richer in vitamin A has been more than compensated for by added pressure on the eyes through various kinds of strain which we shall discuss shortly.

Rasmussen also suggested that nutritional deficiency was compounded by various features of Chinese daily living which caused unusual stress on the eyes. He developed the well-known 'eyestrain' thesis which currently dominates research on myopia. He argued that the effects of malnutrition were increased by 'the strains, pressures and contractions' caused by straining the eyes. The distortions, he argued, 'must take the form of flattening in the horizontal plane, elongation of the part of the axial diameter, contraction or tapering of the ciliary region, with increased curvature and forward displacement of the crystalline lens. The sum of these distortions, varying with each individual, squeezes the optical system away from the retina, lengthening the distance between the retina and the nodal points.'

What then were the causes of the 'strains, pressures and contractions', and is there anything unusual which might suggest that they were worse in China than in other civilisations? Here Rasmussen develops an intriguing chicken and egg argument: the tendency towards myopia led the Chinese to concentrate on close, intricate work, which then made their myopia worse.

He discussed the early development of China as a great literary civilisation with a higher concentration on writing than any other. Close attention to writing and reading put an extra strain on eyes. He drew attention to the early and extensive development of calligraphy and painting, which required 'intensive use of the sight', and to porcelain and cloisonné manufacture 'both arts requiring intensive use of the sight'.

The situation was compounded by the lack of glass windows and of appropriate furniture. Craftsmen worked in ill-lit sheds and in back rooms with hardly any light coming through the oiled paper. The problem was particularly serious in schools. It is thought that myopia tends to develop between the ages of four and ten. Rasmussen used the famous photograph of children crouching over their work and wrote at length about the study habits of children. For example, the 'writer has seen children resting their left cheeks on the left fists while pointing to characters or writing with the right hand scarcely two inches away. It was not uncommon a few years ago in China to see a class of children bent low over their books, the only part of their heads visible to the teacher being the tops.' The 'overconcentration on written characters as a source of learning ... was bound to create eye strain and seriously develop inherent tendencies'. Classrooms 'were not merely poorly illuminated, but in most cases hardly illuminated at all'.

Suppose we take the thesis that intensive education, particularly when combined with a difficult script, is highly significant, what evidence is there for this? One way to proceed would be to look at class and occupational differences. Early work in Europe at the start of the twentieth century suggested much higher rates of myopia in the upper classes, who were presumably more intensively educated. Particularly interesting are two cases. One is the very high incidence of

13. Myopic children in China

Otto Rasmussen, who reproduces this picture in *Chinese Eyesight and Spectacles* (p. 58), describes this as 'the most-published picture in ophthalmology; purchased by the writer in a photographer's shop in Tientsin about 1924 from an amateur's snapshots of "typical scenes in China", neither posed nor commissioned.' The children are writing with their heads very close to the paper. The learning and writing of up to three thousand extremely detailed Chinese characters will produce particularly high levels of eye strain for Chinese influenced populations in eastern Asia. The effort is far greater than for the standardised letters in use in Europe. Often the absence of good light, because the schools had no glass windows, exacerbates the problem.

myopia in parts of Germany. At the start of the twentieth century the figures for myopia among close workers were about 50 per cent in Germany, while it was about 25 per cent in England. Other figures contrasting schools at the same time show a similar difference. It was suggested at the time that the gothic type of Germany was particularly difficult to read and that this was the cause.

Another intriguing case is that of orthodox Jews, made to read, learn and intensively study from a very early age. Recent figures suggest that over 80 per cent of teenage orthodox male students are myopic, three or four times higher rates than the rest of the population. Connections with close work in certain occupations, for instance, the jewel-cutting industry, might be suggested. And an amusing story illustrates the effect of myopia on the ability to play games.

A television programme in July 1999 entitled *The Worst Jewish Football Team in the World* showed a team of under-thirteen-year-old players in the north Manchester league which lost 17–0, 20–0, 23–0 and 25–0. Intriguingly their sponsor was a local optician, who had perhaps got to know many of them quite well. He said that 'he hadn't known when he agreed to sponsor them just how bad they were. When he did see them, his first thought was that they needed their eyes testing, though most of them were already wearing glasses.' One of the young boys suggested the following reason why they were different from other teams: 'They just pass it [the ball] while we're busy standing there or sitting down or whatever.' The optician may well have been right; even with glasses, distant vision may be difficult. Better to sit down and wait until the ball bumps into you.

The connection between close work and myopia has been noted for a very long time. In his *Treatise on the Diseases of Tradesmen*, published in English in 1705, Bernardo Rammazzini discusses the disastrous effects of fine work in industries such as lace-making and sewing in Europe. Moving on nearly two centuries, Browning in a widely reprinted textbook of the late nineteenth century wrote that 'Short-sight is due to two causes: concentrating our attention almost exclusively on near objects – as in reading, drawing, needlework, etc.; and never using our eyes for any length of time in examining objects at a distance. Small-type schoolbooks are most destructive of the sight, especially for very young children.' Subsequent research and specialist studies have confirmed the hypothesis. Clerks, seamstresses and compositors were traditionally highly myopic.

It is not just a matter of small print or bad light. Other factors include posture (sitting at the wrong distance), poor definition, poor contrast, the actual size of objects, poor legibility. Human eyes strain to see what is presented to them. If there is sustained difficulty they become distorted.

A recent overview of myopia edited by Grosvenor and Goss suggests that near work is the most important factor in its development. Particularly important here is education. Myopia is less frequent in populations where there is no compulsory education. Sustained effort to bring small objects, such as letters or numbers, into focus increases the pressure in the eye and accelerates elongation of the eye which produces myopia. Thus one study showed that workers who spent a lot of time looking through microscopes tended to become myopic within two years. It is perhaps this attention to minutely detailed writing over long periods of time that leads to what is known as 'Lawyer's Myopia'. Famously in the film *Some Like It Hot*, Marilyn Monroe made the connection and seeking a wealthy suitor, she hunted down a man who was prematurely wearing glasses as a result of scanning the tiny figures of stock exchange transactions in the newspapers.

Here we come to one of the most intriguing parts of the argument. If attention to detail and close work in bad light over long periods affects the eyes, and if it is more than just simply a matter of direct eyestrain, but also of the connection between the eyes and the brain, in other words the intensity of concentration, we might look at the question of education and writing anew. Western opticians have long suggested that the increasing hours of schoolwork of our children's pressured lives may be behind the rapidly rising rates of myopia (along with other suggestions such as computers, television, etc). But if this is increasingly true in western countries, it is much more so in eastern Asia.

In Japan and South Korea children often go to pre-school and start serious education when they are three or four. When they are in primary school they do very long hours. We summarise a visit to a Korean girls' middle school as follows: 'Visit girls' middle school and are allowed to film a class learning Korean ... Children start school at 8.30 a.m., finish at 4.30 p.m., then go to "crammers" where, in bad light and general noise, they continue to study until 10 p.m. We were told that when they return home they often engage in Internet chat until 2 a.m. Their eyes have about five hours rest.' By the age of seventeen, they might well have extended the cramming period to beyond midnight. There are very few breaks for games, cultural activities or anything else.

We may think that this is a new phenomenon, and it is true that pressures have increased. But it has for very long been a feature of Confucian-influenced cultures where education and learning the classics meant so much. An account of the examination system in nineteenth-century China by Dyer Ball gives an indication of the pressures of intensive education.

In an article on 'Examinations' describing the system of civil service examinations which lasted until 1903, he writes as follows:

> In this strange land there was in vogue for centuries, and even millenniums, a system of examinations, which originally started with the object of testing the ability of these already in office, gradually widened in scope till it became all-embracing in point of geographical extent, and was the test of ability which all had to undergo who desired admission into the Civil Service of the immense empire with its thousands of officials; with this end in view, boys were incited to learn their lessons and be diligent; with this aim, men pursued their weary course of study, year in and year out, till white hairs replaced the black, and the shoulders, which at first merely aped the scholarly stoop, eventually bent beneath the weight of years of toil. No other country in the world presents the curious sight of grandfather, father, and even son, competing at the same time.

He describes how people went on trying to pass until their eighties, with 'untiring perseverance and indefatigable toil'. He continues in the same vein for several more pages.

This is just part of a system which one can still see if one visits a Japanese school, where enormous pressure to memorise classical writings and master a vast literary inheritance through a maze of complex language is placed on young children. Likewise, the knowledge that education is the one gateway to a good job and status in a meritocratic civilisation puts huge pressure on parents and children. The fact that big department stores have areas which specialise in recommending and selling the appropriate clothes for mothers who are taking their tiny infants for an interview at a good kindergarten is just one small indication of this. Another is the famous image of Japanese children sitting past midnight in the cramming establishments, literally holding their eyelids open with matchsticks. But it is not just a matter of the hours of work, the lighting and the parental pressures. There is probably something else which has scarcely been noticed but we think is just as important. This is the nature of what is being learnt, and in particular the writing system.

When asked, Dr Tokoro said that he thought that what was really crucial, apart from the long hours, was the extreme pressure that was caused by trying to write and memorise the three vocabularies which constitute Japanese, in particular the two or three thousand 'kanji' or Chinese characters which are essential even to read a newspaper. This is so difficult that almost half the time that a Japanese child spends in school is devoted to language learning – hence putting pressure on other subjects and lengthening the studying day. The characters are very intricate, have to be accurate (and preferably beautifully executed in a world that values calligraphy so highly), and above all remembered for life.

It seems more than a coincidence that the high myopia belt is concentrated in the places where the Chinese characters are learnt. Singapore, Taiwan, China and Japan are the most extreme cases. South Korea is a very interesting example. The pressure of hours and cramming are almost as high as in Japan. But the Koreans developed a phonetic script (*hangul*), with a small character set, in the fifteenth century and this is now used in all teaching until high school. So language learning constitutes only about one sixth, rather than a half, of the school lessons. One has the school pressure, but one does not have the Chinese characters in the earlier years of school. What are the results in terms of myopia figures?

Although the figures are very small and impressionistic only, they are intriguingly just what one might predict – intermediate between Japan and the West. In two elementary classes (average age 9.5), the proportion of children wearing eyeglasses was between 8 and 12 per cent. In a boys' middle school, with ages between twelve and fourteen, the averages were between 10 and 20 per cent, and in a girls' school, one quarter of the twelve students in one class had glasses. In a class of fifteen-year-olds one-third of the thirty-six pupils were wearing glasses. A little over a third of fifty-four elementary school teachers wore glasses. The general impression from these figures is that the proportions lay exactly between the rates one would find now in Japan and England. Unprompted, the English master said that he believed that children's eyes were getting worse, and all the teachers were hostile to the crammers, but said nothing could be done because the 'biggest problem in Korea' was parental pressure on their children to study hard and get into a good university.

We have seen that nutrition and close work are among the causes of fluctuations in the rate of myopia, and that the extraordinarily high rates of myopia in Japan and Singapore nowadays are likely to be closely related to the educational system, and perhaps the learning of Chinese characters. The high rate of myopia may be one factor in the non-development of optical glass in eastern Asia, until recently. Furthermore, if we consider not just glass but vision in general we can see a potentially interesting difference between civilisations.

One way in which to approach a period when records are meagre is through looking at possible effects. This also has a wider interest, since ultimately it is the consequences of differences in vision that we are concerned with. One of these may be seen in the nature of Chinese art. The suggestion was made by Rasmussen that the fact that Chinese (and Japanese) paintings and drawings are famous for their very minutely detailed foregrounds and vague mountains and clouds in the backgrounds, might have something to do with myopia. Even if there must have been many other contributory causes to this well-known feature, as discussed in the previous chapter, the study of visual representations over the centuries will prove a fruitful area for exploring the history of eyesight.

There does seem to be something intriguing here in the relation between painting and myopia. It is a theme partly picked up by Trevor-Roper who points out that myopic painters necessarily tend to avoid detailed representations of objects outside their limited range. This also affects the distance from which a painting should be seen. To appreciate many Chinese and Japanese works of art one needs to stand much closer to them than to some of the more famous examples of western art. Trevor-Roper even notes the fact that Chinese artists tend to concentrate most of the detail in the lower left triangle of their paintings, though he does not, unlike Rasmussen, link this to myopia, but rather to 'gravity'. The nature of the medium on which the painting is done (absorbent paper) and the artistic medium (water-based inks) are also likely to alter the situation.

It is tempting to go beyond this to other art forms. For example, in Japan one wonders whether any of the conventions of drama – Noh and kabuki – were related to myopia. There are the sounds and gestures of kabuki, the absence of importance of facial expressions (fixed expressions are painted on the faces, instead of allowing them to change subtly as the drama unfolds), the huge costumes, particularly using the red end of the spectrum, the fact that (as can be seen in traditional prints) the audience tend not to look at the stage and often face away from it, even the existence of a projecting footway to take the actors out into the audience, so they can be seen more closely, may be important. All this suggests the audience found it difficult to see what was going on at some distance in a dimly lit hall.

The red and gold kabuki costumes could be evidence that widespread myopia may also have affected the discrimination of colour. Whereas a typical European will see the blue end of the spectrum best, it seems likely that Chinese, Japanese and Koreans see the red end of the spectrum more clearly. Trevor-Roper noted the predominant part played by red and gold in Japanese and Chinese art and that there was no specific word in Chinese for blue. To this could be added that the primary colours in these three countries, apart from black and white, are brown-red, yellow-red and blue-green. That two should be from the red end of the spectrum and the other from the region closest to that end is interesting. Likewise the fact that many ceremonial and dramatic costumes are red and yellow is notable, as are the reds and golds in many temples and royal buildings. Reds are more predominant in Shinto shrines and Chinese temples, though turquoise is also highly favoured in royal palaces. The famous red, rather than yellow, sun on the Japanese flag is also often mentioned. Trevor-Roper explains the link with myopia as follows. 'The blue rays of light are refracted more than the red, and so are brought to a focus slightly in front of the normal retina, and the red rays correspondingly just behind it; hence the myope, with his abnormally long eye, will see red objects in better definition ...'

It would also be interesting to look more closely at literature. Certainly in the west the contrast in the images used by a myopic poet such as Keats, and a normally sighted one such as Shelley, is illuminating. Those of Keats are based much more on the other senses of smell and sound, or are much more fanciful, while Shelley deals with distant prospects. It might well be worth examining the very rich literature of Japan and China with this in mind.

There is also a certain amount of anecdotal or indirect evidence. There is the considerable concern in traditional Chinese medical tracts with eye diseases of various kinds. Older ethnographies record the widespread presence of eye medicine shops, some of which had been in operation for more than seven generations. There is evidence that some ordinary Japanese children had amazingly acute microscopic vision. Edward Morse in the 1870s described how he was showing 'a little country boy' the features of a beetle which 'when placed on its back jumps into the air'. Morse showed him how this worked 'with the aid of a pocket magnifier'. In the west 'only entomologists are familiar with this structure; yet this Japanese country boy knew all about it, and told me it was called a rice-pounder ...'

In the work by Li Yu translated as *The Carnal Prayer Mat*, originally published in 1634, we are told of a lady who was 'nearsighted'. The author explains that this made her especially attractive. 'Most nearsighted women are pretty and intelligent. And for women there is a certain advantage in nearsightedness: it makes them save their feelings for the serious business of marriage instead of squandering them prematurely on passing adventures.' He approvingly quotes the 'popular saying' that 'She may be nearsighted, but in the marriage bed she knows what she is doing.' Thus 'nearsighted women are largely immune to temptations of this kind ... It has been recognized from time immemorial that marriage with a nearsighted woman is usually happy and free from scandal.'

There are many other areas which might provide clues. As Mann and Pirie suggest, to consider the effects of long-sight on 'seating arrangements in cinemas and theatres, the size and position of public notices such as signposts, the use of the blackboard in schools and others of our common arrangements, which presuppose clear sight for things more than six yards away'. If we turn this on its head, and wonder what a civilisation with very high myopia rates might exhibit, a number of curious features of Japanese civilisation take on a new meaning. What must it be like to live in a world where everything beyond about one foot (a third of a metre) is blurred?

So, in relation to Japan, for instance, one is led to wonder about the ceremonious bowing which is more easily observed than the minutiae of facial expression, the generally impassive expressions, the giving of name cards to indicate identity, the emphasis on whole body communication (*hara*) rather than facial gestures or speech. What one would have to do is to walk round Japan with glasses which created myopia – and then see how many things were visible and how human art and gestures and use of noise had made them more so. Is this a landscape created for the visually challenged?

One intriguing area is the non-development in China, where gunpowder was invented, of far-sight weapons. Of course there are other reasons for this, but it may well also be partly related to the fact that the Mandarin class, at least, would not have found them easy to use. A historian of China, Mark Elvin, pointed out to us that deterioration in eyesight might be seen in changes in hunting, shooting and even stargazing. Some might even facetiously point out that it may thus be more than a coincidence that the Zen art of archery required the archer to learn to fire the arrow without having to look at the distant target. After constant practice, the archer intuitively knew how to aim without consciously looking at the target. This would be specifically an activity for the elite.

Another effect might be the skills in 'micro work' which have long impressed western observers of Japan; the intricate traditional crafts of lacquer, *inro* making, *netsuke* carving, bonsai (miniature plant) growing, the subtle movements of the tea ceremony. Many of the professions which involve minute work have in the past attracted people with short-sight; ivory carvers, miniature painters and so on. That this may have a direct continuity into the Japanese skills in certain branches of modern manufacturing, many of which have the suffix 'micro' (micro-engineering, micro-electronics, micro-computing), does not seem implausible. The instructions on how to use the goods manufactured in Japan after the Second World War were also in minuscule characters, practically unreadable to a westerner.

One also wonders about the shape and nature of Japanese

homes and their furnishings. The rooms were tiny, as were the houses, and their simplicity and the absence of furniture would make them ideal for a people who found focusing on objects even at the far end of a large room difficult and who might find moving around crowded furniture trying.

Another well-known feature of Japan is the emphasis on the other senses. The sense of smell is particularly well developed. Not only are the revolting smells of westerners often commented on, but also there has been a huge development of the art of discriminating scents. Numerous books exist on the subject and the palate of different scents and incense was vast. The *Genji*, an eleventh-century novel, is full of competitions to detect subtle variations in scent and the approach of the Prince was often heralded by his perfumes.

Likewise there is a very great emphasis on sound. Tiny sounds are noted and used: the sound of water dripping, the tinkling of bells, the sound of a frog jumping into a pool, various taps and clicks during the tea ceremony. A Japanese friend suggested to us that the Japanese had a better sense of hearing to compensate for their poor eyesight.

Again, one notices the love of whole masses at a slight distance, rather than small things which might not be seen. Thus the cherry blossom tree, a cloud of pink or white, a full moon, the shape of Mount Fuji which is so distinctive and obvious. Mann and Pirie poignantly capture this as follows: 'If a child is born short-sighted it will make no complaint. This is often not realized, but is obvious if we stop to think ... a small short-sighted child given glasses for the first time and sent for a walk shows the state of mind. She said "Look! Do you know that a tree is made of leaves?"'

There is a notable absence of 'prospects', of viewing

points, of sightseeing spots, in traditional Japan. This was noted by Timon Screech, who suggested it was because of an attempt to stop the population from spying out the lie of the land, but this can hardly explain it. It is much more likely that many people would not have climbed the crags or towers which filled western countries since they could not see from them until the introduction of eyeglasses from the west. After the use of eyeglasses spread in the eighteenth century, going up in balloons and looking out over the countryside became a craze in Japan.

The last area is the most controversial, but it is worth considering. This is the effect of myopia on personality. This is difficult ground, but here are a few suggestions. Mann and Pirie suggested that a myopic child is

> intensely interested in books and in all fine detail and very bored with games ... Such children win scholarships and may be correspondingly unpopular ... they drift into jobs entailing close work, they get round-shouldered and peer at their work. They get wrinkles at the corners of their eyes because they are always screwing up their lids to try to make a sort of 'pinhole' camera to help them to see distant objects better, and this gives them a feeling of strain.

This theme of the bookish myope, with a card-index type of memory, very highly aware of colour which is used for distant cues, inward-looking, finding extrovert behaviour difficult, bad at team games and so on is to be found in much of the literature.

The theme has been explored by Trevor-Roper in a section called 'The Myopic Personality', which takes as its theme an eye-specialist's remark: 'But you don't understand, we myopes are different people.' He refers to 'the studious and rather withdrawn myope', and quotes a long and fascinating passage from a Dr Rice on these themes.

> A near-sighted child cannot do well on the playground because he cannot see. He will not like to hunt because he cannot see the game or the sights of his gun. He will not like to tramp because distant objects are poorly seen and, for that reason, not appreciated. He will not like races or aviation or travel or sports of any sort. As a rule these persons do not like the theatre, or the motion picture ... The child who knows that he cannot excel over his fellows in games gets a big satisfaction out of the conquest of the mind that he can command ... He pleases his teacher but he loses his friends ... Such a child as we have described is not dependent on others for entertainment and is liable to grow rather contemptuous of the abilities of others. He does not adapt himself to the surroundings and is not willing to make compromises.

In any discussion of biological and other differences, it is extremely important to avoid any overtone of moralising or superiority. Even the title of a chapter along the lines of Trevor-Roper's *The World Through Blunted Sight* would be inappropriate. It is true that myopia is now regarded as a disability and many consider it a misfortune. In fact, it has many advantages and has been behind much of the creative work in both east and west. A person with myopia probably inhabits a more intense, intimate, meaningful, kind of world. Myopes see the world at a different angle, in an unusual perspective, which is perhaps why myopes are more highly represented even in art schools and, when they are there, prefer not to have their vision corrected.

Many of the greatest poets have been myopic: Milton, Pope, Goethe, Keats, Tennyson, Yeats. Among writers, James Joyce and Edward Lear were myopic. Notable myopic musicians included Bach, Beethoven, Schubert and Wagner. Gregor Mendel, the first investigator of modern genetics, was another myope. We are told that among really gifted mathematicians, myopes are four times as common as in the general population.

Perhaps more surprisingly, many of the greatest painters, for example Van Eyck, Dürer and possibly Vermeer, were myopic. Myopia was particularly prevalent among nineteenth-century impressionists, for example Cezanne, Degas and Pissarro. In a twentieth-century French art school, the proportion of myopes was about twice the national average. Myopes are successful in education and the arts. They also have the compensation of having less of a problem in reading without glasses in older age.

It is thus essential to avoid the pejorative overtones reminiscent of what one used to find in school playgrounds, where children who wore glasses for myopia were picked on as swots. It would be very unfortunate if it were to be thought that one was arguing that the half of the world's population who use Chinese characters (not to mention orthodox Jews, Indian Brahmins and other more than usually myopic sub-groups) were somehow inferior to the long-sighted peoples of the western world. On the other hand, it is not wise just to ignore the possible differences for fear of being thought politically incorrect, racist, Orientalist, determinist or whatever. Instead we can see that there is a possible added twist to the story of the development of glass instruments in east and west.

In relation to the high-level civilisations of Eurasia (apart from Islam, which is another story), in other words China and Japan, there may have been a movement towards a world with an attention to microscopic detail, the dominance of senses other than the eye, combined with less accurate long-sight. If one wants to push the thesis to its extreme, one could suggest that this helps to bring out some of what is peculiar about the west. As we shall see, western glass technology extended the already reasonable eyes with telescopes and microscopes, and corrected the effect of old age with spectacles. The Chinese and Japanese turned their eyes into virtual microscopes, having earlier had eyes which were virtual telescopes. But this was at some cost to themselves and helps explain some unusual features of their civilisation, just as glass does in the west.

In China, Japan and for a time Korea, eyesight, or at least the ability to see far off, may actually have deteriorated. Reliable knowledge may have levelled off or even decreased. The world became flatter and closer, the authority retained from the past when the world seemed clearer was increased, curiosity to discover new things was diminished, the written word and memory of past achievements were stressed. In the west, seeing is believing; in the east, hearing and reading is believing. The senses of smell and of hearing and remembering are given greater weight in the east; experimenting and touching and seeing in the west. This was reinforced by language, a veil of complex writing in the east, whereas the west increasingly became an oral and visual culture. Both have their charms and their advantages, but in the grim world of practical politics it was the western solution, to extend the human eye with spectacles, which won the competitive battle. Now the entire world has spectacles. And we have forgotten the thousand years of history when things were very different.

9

Visions of the World

Churches, Palaces, Castles, and Particular Houses, owe their chiefest Ornaments as well as Conveniences, to Glass; for that transparent Substance guards them within from too great Heat and Cold, without hindring the Intromission of the Light. Looking-Glasses, and other great Plates of Glasses are so many surprizing Objects to our Eyes, representing so distinctly and naturally all even from the least to the greatest Actions of the Objects before them, whereby also one may always keep himself in a neat and agreeable dress. Notwithstanding not one in a Thousand of those who have them, ever reflect on the Admirableness of the Work, which is beyond doubt, one of the chiefest, and most perfect Pieces of Art, and than which Man Can make nothing more wonderful.

Haudiquer de Blancourt, quoted in Raymond McGrath and A. C. Frost, *Glass in Architecture and Decoration* (1961, p. 5)

THERE IS A GREAT MYSTERY concerning how our modern world emerged. If we ask what have been the half dozen greatest transformations in the field of human knowledge, both in content and form, there can be little doubt that the Renaissance and the Scientific Revolution would be among those selected. Along with the evolution of language and discovery of writing, the transformation in the tools to help humans understand nature that occurred between about AD 1300 and 1800 is clearly of enormous importance. Upon them have been built the new technology of industrialism, the new social system, the communications networks, the new political system and the global culture which we now experience. Looking at human history as a whole, the effects of what has happened in the very recent past, and which first happened in only one corner of the world, are dramatic indeed. Yet when we ask the question why did one of the great changes in human history occur, it is difficult to find any satisfactory answers.

In order to solve this mystery, we first need to define the question more carefully. What exactly do we mean by the terms 'Scientific Revolution' and 'Renaissance'? The 'Scientific Revolution' needs, in fact, to be split into two 'revolutions'. The earlier one occurred roughly between 1250 and 1400 and it consisted of several features. These included the absorption of Greek learning by way of Arabic scholars, the development of universities, the improvement of logical tools, a growing concern for precision and accuracy, the increasing sophistication of mathematics, chemistry, physics and in particular optics, a stronger emphasis on the authority of observed visual evidence rather than the authority of the ancients as written in texts. This first revolution laid the necessary foundations, including the experimental method and method of scepticism, doubt and suspended judgement, for the more famous scientific revolution, which is usually dated from the 1590s through to the end of the seventeenth century. This second scientific revolution explicitly laid out the scientific project as we know it, with its use of scientific instruments in order to gain very large quantities of new reliable knowledge.

If we broaden the definition of the scientific revolution in this way we can see that it was not just a sudden breakthrough, but had its roots deep in classical thought, which, combined with Arabic advances, flowered in the work of medieval thinkers. We are thus dealing with something which covers over half a millennium, from about 1200 to beyond 1700. Likewise the geographical range is narrow for while many elements were found elsewhere, the whole set of interconnected parts was, for a period, uniquely found only in areas of western Europe.

This is the puzzle. Why did this great event which transformed human vision and understanding happen then (1200–1700), there (parts of western Europe) or at all? There was clearly nothing inevitable about it. Indeed, greater civilisations with more sophisticated technologies and social structures showed few signs of having such a revolution. It was clearly an event which changed our world, so why did it happen?

If we turn to what we call the Renaissance, it is conventionally thought of as occurring in the arts (painting, architecture, literature) and to have a set of features which include a growing precision of observation and representation, the mathematisation of the rules of painting and architecture, the development of methods to represent perspective so that three-dimensional space could be convincingly depicted on a two-dimensional surface, growing realism in the portrayal of nature, new architectural and poetic devices which gave increased intensity and power, new concepts of the individual and his or her place in the universe, and a new concept of time.

Once these characteristics are listed, we can easily see how this set of attributes overlaps greatly with the scientific or knowledge revolution. This overlap is nowhere better represented than in the work of Leonardo da Vinci, both a 'scientific' and a 'Renaissance' genius. It is easy to see that both movements are basically about the extension of reliable knowledge. Gains in one field, for example mathematics or the representation of three-dimensional space, soon feed back into the other. The most obvious example of this is in the field of optics, which was the foundation discipline in both the early scientific revolution and Renaissance art.

The Renaissance in these senses started in the same period as the first scientific revolution, that is about the middle of the thirteenth century, and it extended up to about the start of the second phase of the scientific revolution, that is up to about 1600. The area in which it occurred was roughly the same, with a particular locus in northern Italy and north-western Europe. It was not to be found in any civilisation outside western Europe.

Since the scientific revolutions and the Renaissance were really all part of one phenomenon, manifestations of one tremendous development, so it is sensible to stop differentiating them, just as it is profitable to suspend the distinction between the first and second scientific revolutions: both are encompassed within what we might call a knowledge revolution.

Having done this, we can see that in order to explain why the knowledge revolution occurred, any explanation has to meet certain criteria. Since the rate at which reliable knowledge was produced began to increase markedly from the thirteenth century, an explanation must contain elements which were present from that date. Hence the invention of the printing press or the discovery of the New World, both occurring in the fifteenth century, are too late to be among the precipitating events, though they may sustain and expand the movement. Then again, an explanation must suggest factors which grew rapidly after about 1200, being largely absent or muted before. Furthermore, whatever the factors, they must be particularly strongly present in both Italy and north-western Europe, for the knowledge revolution occurred simultaneously in these two areas. Furthermore, using the comparative method, the factor or factors must be largely absent in all other civilisations, where the knowledge revolution did not occur at this time. Finally, co-existence or coincidence is not enough. It must be possible to show how the factor or factors actually produced the central feature of the knowledge revolution. That is to say, how could it or they have directly or indirectly encouraged a more precise, realistic and detailed knowledge of nature, helped give the basis for the rules which made the representation of nature more accurate, and stimulated curiosity and confidence in the pursuit of this objective?

Judged by these demanding criteria we can go through the various explanations which have been advanced by numerous writers to solve these puzzles in order to see how far they meet these tests. If we leave on one side for the moment the deeper philosophical and cultural factors which provided an indispensable basis for what happened, it is worth listing the most likely candidates which could be contenders for an explanation. These include the following: the mechanisation of the world view through the development of machines; specific legal traditions; the growth of particular types of city; a particular social structure; trade and exploration; a plural and yet culturally united civilisation; the development of commercial capitalism; the development of logical and rhetorical methods; the improvement of methods of storage and dissemination of information (e.g. printing); universities; tools of time (clocks); arts of memory; tools of measurement and calculation; networks of knowledge; trusts and associations allowing collaborative long-term work.

Clearly all of these are important. Yet we can short-circuit the discussion by stating that few of them seem to meet all of the criteria set out above. Probably the three most promising are mechanical clocks, universities and other corporate institutions, and a particular fragmented yet unified political and economic system. Yet there still seems to be something missing. We have a picture of many of the pre-conditions, yet what tied them together and allowed one civilisation to move toward new systems of thought still seems to elude us. So where else should we look? If we were detectives, we might look for something that has been overlooked because it is too obvious, staring us in the face. We believe that an obvious missing factor in our understanding of how one civilisation broke through into a high level of reliable knowledge, is glass.

We have argued that glass transformed humankind's relation with the natural world. It changed the sense of reality, privileging sight over memory, suggested new concepts of proof and evidence, altered human concepts of self and identity. The shock of the new vision destabilised conventional wisdom, and the more precise and accurate vision provided the foundations for European domination over the whole world during the next centuries.

The story of what happened now seems relatively clear. In central and eastern Eurasia the shape of the history of glass is basically uniform. The knowledge of the substance and how to make it spread from its source in the Middle East, probably by at least 500 BC in the cases of India and China. The revolutionary new method of glass-blowing was known about AD 500 at the latest. In India practically no glass industry developed as a result of this except for bead and bangle manufacture. Likewise in China and Japan glass was seen as a cheaper, but inferior, way to make decorations for secular and religious purposes. Glass manufacture reached a high point in Japan in the eighth century and then faded to nothing and by 1500 the art was lost. In China it expanded until about the tenth or eleventh century but again faded away after about the twelfth century. Islamic civilisation was a bit different, lying exactly between the two extremes. Glass-making flourished and by the eleventh century Syria and adjacent areas were the most sophisticated glass-making centres in the world. But then glass manufacture declined abruptly and hardly any glass of any quality was made between 1400 and 1750.

In western Eurasia, the history of glass was different. In the south, Roman civilisation gave the impetus to wonderful domestic glass for utensils, especially for wine, and hence to Venetian mirrors and wine glasses. In the north, Christianity plus the climate were conditions that encouraged flat and coloured glass in the medieval period. So the glass technology improved rapidly, spurred by luxury, trade, religious zeal and the desire for comfort. Early on, the skills and knowledge of mathematics were good enough to make spectacles and the desire to overcome presbyopia made this popular. With lenses, prisms and spectacles a growing interest in the properties of light could be explored which later led through to microscopes and telescopes. Likewise the development of glass artefacts and of mirrors had deep effects on chemistry and astronomy. The effects on the perception of space, on the dominance of vision, on health and agriculture were considerable.

We can examine this story in a little more detail by dividing glass up into its major uses. The use of glass for 'verroterie', that is glass beads, counters, toys and jewellery, is almost universal, at least in Eurasia, though even this was absent in the half of the historical world comprising the Americas, sub-Saharan Africa and Australasia. This is really the exclusive use for glass in India, China and Japan over most of the last two thousand years. For this purpose, glass-blowing is not absolutely required, nor does this use have much influence on thought or society; its influence is in the field of luxury goods and aesthetics. Basically glass is a substitute for precious stones. Hardly any of the potential of glass as an instrument for knowledge or for improving the physical environment is exploited.

Historically the use of glass for 'verrerie', that is vessels, vases and other containers, was largely restricted to the western end of Eurasia. There was very little use of glass for vessels in India, China and Japan. Even in the Islamic territories and Russia, the use declined drastically from about the fourteenth century with the Mongol incursions. In relation to China, in particular, this use can be seen as mainly an alternative to pottery and porcelain. The great developers were the Italians, first the Romans, with their extensive use of glass, and then the Venetians with their 'cristallo'. Much of the technical improvement of glass manufacture arose from this and it is particularly associated with wine drinking. Thus we have a phenomenon much more specific in scope, finding its epicentres in Italy and Bohemia. There are various links to science here, for example the fact that the fine glass needed for the earliest microscopes was made from fragments of Venetian wine 'cristallo'. Likewise the development of tubes, retorts and measuring flasks for chemistry, as well as thermometers and barometers, developed out of this.

The use of glass for 'vitrail' or 'vitrage', that is window glass, has also been restricted until very recently, but to a slightly different area. Historically, window glass was only found at the western end of Eurasia, and China, Japan and India hardly developed it. More surprisingly, perhaps, nor did the great 'verrerie' area of the Mediterranean, Islamic and Roman areas. Although they knew of the possibilities, the glass window was little developed. The great window revolution mainly occurred in Europe north of the Alps. Two of the main factors were cold climate and religious architecture, incorporating the Gothic stained glass window. From about the eleventh century stained and then domestic glass, with its attendant technological developments, spread and transformed architecture, social life and thought, but only on a large scale in north-western Europe.

A fourth use of glass comes from its reflective capacity when silvered. The use of good glass mirrors was again circumscribed in time and space. The development of glass mirrors covered the whole of western Europe, but largely excluded Islamic civilisation, perhaps for religious reasons. Glass mirrors were also not developed in India, China and Japan. Nor were they really developed by the Romans. They are thus temporally contained, only becoming common and of high quality from the thirteenth century, and geographically limited to western Europe. Yet they are a crucial feature in the development of the sciences of optics and of perspective in art. Without them, much of what has happened in the increase in reliable knowledge of the natural world to which we give the terms the Renaissance and the Scientific Revolution would not have occurred.

A final major use is for lenses and prisms and in particular their application to human sight in the form of spectacles. Again, while the concept of the light-bending and magnifying properties of glass were probably known to all Eurasian civilisations, only in one did the practice of making lenses really develop, that is in western Europe. As with mirrors, the developments also occurred quite late, mainly from the thirteenth century onwards. This coincides precisely with the medieval growth in optics and mathematics, which fed later into all branches of knowledge, including architecture and painting. It also influenced a specific and important sub-branch of lenses, the development of spectacles. Glass spectacles were not used in Japan, China, India, Rome or Islam. Only in western Europe from about 1280 onwards did they begin to spread, later to form the crucial step to microscopes and telescopes.

From this we can see that the more than half of the Eurasian population, which was constituted by India, China and Japan, only had glass for one of the five purposes. The middle section, Russia and Islam, added some use of glass for vessels. Western Europe as a whole made use of glass in four ways, by adding mirrors and lenses, but only from the thirteenth century. North-western Europe had all five major uses in profusion by adding windows.

We believe that there is more than a coincidence between some of the major divergences in the knowledge systems of civilisation and the development of glass. Firstly, there is the correlation in space. The area where glass developed in multiple ways, western Europe, was the area where a new world vision roughly lumped under the terms 'Renaissance' and 'Scientific Revolution' occurred. Despite the fact that Islamic and Chinese knowledge was far more extensive up to the twelfth century, it was not in those areas that the breakthrough occurred. Secondly, there is the coincidence of timing. The rapid development of glass, particularly for windows, mirrors and lenses, occurred in western Europe from the thirteenth century, and this is just the period when the major breakthroughs in optics and mathematics and perspective began to be noticeable. Meanwhile, Islamic civilisation, which had been well in advance, began to give up glass from the thirteenth century and abandoned it almost entirely from the later fourteenth; scientific thought withered away in that area. In reverse, when scientific glass instruments were introduced on a large scale into Japan in the later nineteenth century they led to major developments in technology and science.

If this were just a coincidence in space and time, with no plausible causal connection, we might dismiss it as just a curious parallel growth. But it is not difficult to see the actual effective links. It is apparent in the lives and works of most of the major figures in the west who developed the new world views. There were Alhazen, Roger Bacon and Robert Grosseteste with their explicit use of glass instruments in the development of mathematics and optics. There were the great Renaissance experimenters in perspective and the more accurate observation and representation of nature, from Brunelleschi and Alberti, through Leonardo and Dürer. All used mirrors, flat planes of glass and lenses to experiment with vision and light. Then there were the great seventeenth-century scientists, Galileo, Kepler, Newton and others whose work centred on the investigation of nature through glass instruments. Almost every great scientific advance needed glass at some stage. Furthermore glass helped to extend the most powerful of human organs, the eye. This only happened in one part of the world.

We are, of course, naturally wary of all single-factor and reductionist explanations. It would be ridiculous to argue that glass was the only thing needed. As we have seen, its use is dependent on the context and there were many other factors which also determined the massive increase in reliable knowledge which is the foundation of our world. It was at the most a necessary condition, but not sufficient in itself. Yet it does not seem too much to argue that if one had to pick one factor above all others, more important than the growth of cities, the revival of ancient learning, clocks or printing, then it would have to be glass. Without its development, it is difficult to see how the new world vision could have been established.

Yet this does not mean that there was anything inevitable about the outcome, or that there was any particular design or purpose involved. Glass was developed for other uses; it was its beauty and utility which recommended it to people. Only by a set of giant accidents was this substance also one which would bend light in a way which would change human vision of the world.

Accidental though it was, the effects of this technology were immense, but originally only in one area of the world. The story of glass again shows the very different effects of a new technology in east and west. Like gunpowder, printing and clocks, which had little or no revolutionary effect in central and eastern Eurasia, glass revolutionised western Eurasia. What, in summary, were these effects?

There is the dramatic transformation of a basically aural, text-based culture, similar to that in all other civilisations, to a visually dominated one. The sense that what really mattered was sight. Here we have argued that the increase in the power of the human eye and mind through the development of glass technologies is one of the most important influences, without which this would not have happened. Though it cannot be absolutely proved, it looks highly probable that glass was one of the most important factors in the peculiar development of a visual, experimental, rationalistic, 'scientific' and realistic world. That disenchanted world which we associate with Descartes and Newton grew out of glass. Glass was absolutely essential for the revolutionary developments in the generation of reliable knowledge between the thirteenth and eighteenth centuries upon which our world is founded.

We hope to have sketched out a possible story of the way in which increased knowledge feeds into better tools of knowledge, which in turn feed back into more knowledge. This helps to solve the long-debated problem of why it was Europe, and not Islam or China, that made the crucial breakthrough to a new and more reliable understanding of the natural world.

Glass is not just a tool to think with, but also a tool to improve comfort and efficiency. The period between the thirteenth and eighteenth centuries in Europe saw many of these potentialities unfold and they are an important part of the story of the intellectual effects. As we have already seen, the intellectual and the material are interlinked. Many of the ways in which glass began to embed increased reliable knowledge in shaping humankind's artefactual world then fed back into increasing the possibilities of further rapid advances in reliable knowledge.

Just as it improved comfort and the length of the working day through windows, glass probably affected health. Glass lets light into interiors and is a hard and cleanable surface. This was one of its attractions to the fastidious Romans in relation to utensils, and likewise for one of the great glass-using and representing civilisations, the Dutch. With their enormous windows, it was in the Netherlands that the use of glass developed most. Transparent glass lets in light so house dirt becomes apparent. The glass itself must be clean to be effective. So glass, both from its nature and the effects it has, is favourable to hygiene. That the two major glass-using civilisations of the seventeenth century, Holland and England, should be widely noted for their cleanliness and good health seems to be linked. It is true Japanese houses achieved even greater cleanliness by other methods and without glass. But in colder northern Europe windows were probably a very important factor for the warmth they provided.

The new substance did not merely alter the private home, but in due course transformed the growing consumer society. Here the focus shifts northwards to England and a century later. The lead glass sheets produced by using coal were ideal for a nation of shopkeepers to glaze their shop fronts with and foreigners marvelled at the results in the eighteenth century. The change was well captured by a French visitor to England. 'What we do not on the whole have in France,' he notes, 'is glass like this, generally very fine and very clear. The shops are surrounded with it and usually the merchandise is arranged behind it, which keeps the dust off, while still displaying the goods to passers-by, presenting a fine sight in every direction.'

As well as houses and shops, the new application began to transform agriculture and knowledge about plants. The use of glass in horticulture was not an invention of the early modern Europeans. The Romans had used forcing houses and protected their grapes with glass. This Roman idea was revived in the later Middle Ages, from about the fourteenth century, when glass pavilions for growing flowers and later fruit and vegetables begin to be noticed. As glass became cheaper and particularly flat window glass improved in quality, the development began to exceed the Roman use. The growing of orange trees under glass was noted in 1619 and a heated glasshouse was built in 1684 in the Apothecaries' Garden at Chelsea. As this happened glass cloches and greenhouses improved the cultivation of fruit and vegetables, bringing a healthier diet to the population. Just as the glass window lengthened the working day for the humans, so it did for plants, changing, as it were, the climate and using solar energy to grow nutritious food for humans. A transformation which is now happening as a result of plastic in many cold, dry and windy parts of the world such as northern China, happened in another way much earlier with glass.

Finally we can note a plethora of other useful inventions which altered material life. Among those that have been noted are storm-proof lanterns, enclosed coaches, watch-glasses, lighthouses and street lighting. Thus travel and navigation were improved. How effective would the successors to Harrison's chronometers have been without glass? Or again there is the effect of glass bottles, which increasingly revolutionised distribution and storage. For example, glass bottles created a revolution in drinking habits by allowing wine and beers to be more easily stored and transported. Since both of these drinks with their tannin and hops were medically very important, the effects may again not only have been to encourage manufacture, trade and agriculture, but also to improve the health of people who could more easily avoid drinking polluted water. The ways in which glass altered the flexibility of storage and distribution is a revolution similar to that caused when freezing and canning opened up new possibilities in the second half of the nineteenth century.

Thus, at first through drinking vessels and windows, then through lanterns, lighthouses and greenhouses and later through cameras, television and many other artefacts our modern world built round glass has emerged. Through another chain of events it revolutionised health. Microscopes made the discovery of bacteria possible, the germ theory that emerged led to the conquest of much infectious disease. Glass even affected what humans believed (stained glass) and how they perceived themselves (mirrors). So it entered human civilisation at all sorts of angles, but at first only in one part of the world. These different aspects were also all interconnected in complex ways. For example, windows improved the workshops, spectacles lengthened the working life, stained glass added to the fascination and mystery of light and hence a desire to study optics. It is this rich set of inter-connections of this largely invisible substance which makes it so powerful and fascinating.

This all now seems so obvious. Yet if we look through what has been written on the origins of modern thought or the role of glass the connection seems to have been largely overlooked. So why has glass been so invisible and its social history so little studied? Specifically, why has it not been realised that it is a crucial part of the answer to the largest question in intellectual history, namely why parts of the world witnessed the knowledge revolution of the fourteenth to seventeenth centuries, comprising the Renaissance and Scientific Revolution? Assuming that there is indeed a connection, the omission is important for it leads us to reflect on the methodological approaches needed to study a phenomenon like glass. The factors which are specific to the peculiar qualities of glass itself are discussed in the introduction to this book. Here we will briefly summarise some of the methodological considerations which would affect the study of this and many phenomena in the past.

We believe that one powerful influence on the way in which we have seen the subject has been the effect of combining history with anthropology. Anthropology is a broadly comparative discipline: it concerns all parts of the globe and investigating what is common and what unique about particular institutions or societies. We are constantly trying to detect absences, to look at co-variations, to test the strength of causal links by looking to see what seems invariably to fit together. This is the essence of anthropology and it is what we have done in this book by looking at five different civilisations.

This comparative method also leads us to notice things which are ubiquitous in our own environment. By setting our own against other civilisations where glass is absent we have a backdrop against which its oddness becomes apparent.

Would we ever have 'seen' glass and its importance if we had remained as historians of a particular European country, or even Europe as a whole? One has to go outside the whole system to see something so obvious. If one stares at a phenomenon straight on, one often looks right through it. By altering the angle of vision, suddenly new and important areas stand out.

Many phenomena, including almost all increases in reliable knowledge, can only be identified if they are seen as the result of the work of a network of interconnected centres spread widely apart. Glass technologies in western Eurasia moved from place to place and the whole area from Syria and Egypt to Scotland and Scandinavia has been one intersecting system of people and ideas. In this many-centred, diverse, competitive system lay the secret of what happened and why. This could not have been seen if we had confined our interest to one country.

The anthropological perspective has a considerable time depth. Anthropologists have tried to trace the whole evolution of *homo sapiens* from his ape ancestors up to the present. A thousand years is quite a short period from this perspective and its effect is to make us consider the last hundred thousand years as a whole. We then get a sense of long-term development and widening trends. This wide historical frame allows us to put events such as the Knowledge Revolution (in science and art) into perspective. We can investigate before, during and after the event. It is then much easier to pick out long-distance connections, to see 'buried' links, things that run underground, or are important for further development but lie well back in the past. A relatively short-term example would be the fact that to understand how the steam engine was developed we need to move back from the eighteenth century through glass used in the discovery of the vacuum in the seventeenth century and then back into medieval glass-blowing.

Anthropology usually has a functionalist approach. It asks what different institutions or technologies do for societies and whether these could be performed by other institutions. This is part of its comparative method, comparing different phenomena in different societies and sometimes showing they have recognisably similar functions. This comparative approach is vital in understanding the history of a phenomenon such as glass: its non-development in eastern Eurasia was not due to lack of knowledge or rationality, but because there were other things which performed the functions which glass performed in western Europe.

The anthropological perspective seeks to be holistic, that is to say it treats phenomena as complex, integrated systems. For example, an anthropologist characteristically studies a particular tribe, village or other group in all its aspects, religious, political, economic, social, aesthetic. The strong barriers between these spheres created by the division of labour, in its widest sense, with the development of 'modernity' are inappropriate in most of the areas where anthropologists traditionally work. In the case of glass, this holistic approach encourages us to see the interconnections between the different features of the past. The conventional distinctions which are useful starting points for thought soon become reified and in the end block investigation. For example, the strong distinction between science and art differentiates the study of the Artistic Renaissance and the Scientific Revolution, or different types of scientific advance, for instance those in the thirteenth and seventeenth centuries. By treating phenomena holistically, it is possible to see that we are dealing with an immensely complex bundle of interconnected features, which includes technology as much as it does religion and which can only be studied if we forget disciplinary boundaries.

Anthropology is also what we might call materialist. There has been an increasing tendency in thought, at least since Descartes' work in the mid-seventeenth century, to separate the material and physical, the province of the natural sciences, from the intellectual and social. This is another division which is challenged by the anthropological experience. Anthropologists often work with living peoples and find it hard to forget the fact, which is often obscured when we only deal with written records, that the physical world is not separate from the mental. Artefacts and technologies have always been a central concern of anthropologists. They even tend to divide their subject by technological criteria, for instance by modes of production and tools. They have often collected artefacts for museums and put on displays to show the relations between material objects and social concepts. So an anthropologist would not be surprised to find that something as physical as glass could alter our world and a number of leading anthropologists have written illuminatingly on the role of various technologies in social variation.

This can be put in a slightly different way. It is very difficult for us to appreciate the way in which the material and the intellectual are interconnected. In trying to understand this we have developed the idea that much of social development can be understood as a triangular movement. There is an increase in theoretical understanding, reliable knowledge of some kind. This is then embedded in improved or new physical artefacts. These artefacts, if they are useful and in demand and relatively easy to produce, are disseminated in huge quantities. This then changes the conditions of life and may well feed back into the possibilities of further theoretical exploration. This triangle has occurred in many spheres of life and the speed of moving round this closed circuit and its repetition lies behind much of what we describe as human development.

The history of glass is an excellent example of this movement between the material and the theoretical which occurs again and again. For example, the improvement in theory (mathematics and optics) led to the development of improved lenses and mirrors, which were multiplied and then fed back into further theoretical developments, which led back into microscopes and telescopes, which later improved health and agriculture and allowed even more research.

In fact, it becomes difficult to distinguish the material and theoretical. Anthropologists have long seen technology as a mix of things and ideas, of ideas embedded or congealed in objects which themselves only have their power from the practices which dictate their use. Thus technology is often defined, as by Marcel Mauss, as 'traditional effective action'. It consists of ways of understanding and changing the world which include things and ideas. Nowhere is this more obvious than in the simultaneous development of ideas and techniques in the making of glass. It is both a tool of thought and a tool with thought embedded in it. What is peculiar about it, is that it is the only substance which directly influences the way in which humans see their world. It is the only substance which is a real extension of a human sense organ, and the most powerful one, the eye.

The anthropological approach to understanding the world is what might be called 'structuralist'. Anthropologists have focused much less than historians on individual people, events or things which are important, but on their relations, on the balances and timing of the forces acting upon them. Thus it is not just the presence or absence of glass we consider, but how much there is of it, how it is used, how it enters into the relations between humans and the natural world, and how it fits with other causal factors which equally need to be considered. This is often combined with a dialectical method which sees forces in an ever-restless movement through a set of oppositions, contradictions and resolutions, as in the famous dialectic of thesis, antithesis and synthesis. Anthropology emphasises social structures and the way in which people do act in concert but are deeply influenced by their social networks, and looks at the degree to which their culture encourages their activities. The development of reliable knowledge is seen not to proceed through a number of steps on a ladder to which we can attach famous names – Alhazen, Grosseteste, Leonardo, Kepler, Newton, Einstein, etc. Rather, these names are mnemonic devices for us, instances of and catalysts in much wider movements of thought.

Furthermore, although not, of course, confined to anthropology, the working experience of trying to understand numerous societies and civilisations reminds anthropologists that causal paths are very complex. It is not enough to use a simple-minded idea that every effect has only one cause, or that cause and effect have to be close in time and space. Once we are aware of the extended chains of causation it is easy to see that if one step is missing, even if it is far back along the chain, then the final outcome will be different. We saw the importance of this in relation to glass when we noted how many times glass was needed at some stage in the development leading to a discovery, even if it was not the immediately enabling factor. Only a consideration of these complex pathways allows us to see the indirect, partly hidden, but powerful influences of something as diffuse and complex as glass.

Looking back over history and using mainly written records, we are often impressed by the purposive, planned, rational, goal-directed nature of human life. From this it is easy to slip into an unexamined form of teleological thinking, to believe that the most important developments are planned, designed by human actors (or by God). The consideration of human civilisations over long periods and in diverse interactions, as well as the microscopic investigation of daily life where humans patently have little idea of the actual consequences of their actions, reminds anthropologists of the importance of unintended consequences. The importance of the random, chance, variation, becomes obvious.

This helps us to appreciate more easily that the history of glass appears to have been largely a set of happy accidents and unintended effects. What happened conforms very closely to a Darwinian, selectionist model. The whole story is an illustration of 'random variation and selective retention'. Things invented for one purpose are then used for others. Indeed, this is the single most important fact to emerge from the history of glass. It was developed to make beautiful and useful things for humans. Only through a giant accident did it turn out that this magical substance could also be used to extend human vision and hence alter thought. So that, in its absence, our modern, affluent civilisation could not exist. We might, as a friend put it, call this the Eeyore Effect. The empty honey pot which so disappointed Eeyore when he received it from Pooh became transformed into something marvellous by the burst balloon from Piglet – a 'Useful Pot for Putting Things In' (and appropriately enough, another use of glass). Anthropologists notice this Eeyore Effect daily. They see that a newly introduced tool or technique – steel axes, a new crop, irrigation system, weapon, plastic buckets, cricket – is very often exploited in many ways which were not originally envisaged.

Furthermore, a new technology can transform a culture well outside the area in which it operates, enabling a much wider array of new things to occur. This helps anthropologists to understand the cumulative nature of technological development, what could be called the 'Meccano effect' because it is like adding a new piece to a set of that famous building kit.

It is a general principle that as each piece of reliable knowledge was added, for example the techniques of glass-blowing, of making fine mirrors or flint glass, this did not merely add one more item to the stock of things humans can do. In fact, it led to the possibility of doing dozens of new things. Just as adding a wheel to a Meccano set transforms the potentials of all the previous pieces, so it was with glass. The effect, unless stopped, is that reliable knowledge and effective action will expand exponentially. This has been the story of the vast growth of the last three hundred years, where human understanding and control of nature has grown at a far greater than linear rate. The history of glass is a very good example of this. Its power and effects have become greater and faster and glass itself has not just been one added resource for humans, but permitted innovation in so many other technologies. As the technology grew more powerful it did not just lead to better drinking glasses, or mirrors, or window panes, but altered health, housing, thought, communications, travel, shopping and a host of other areas.

Putting the argument negatively brings out its strength. Innovators, whether they are innovators of new knowledge or, in particular, new knowledge that is built on the use of artefacts (such as glass), have to use what is available. And if the knowledge and artefacts are not available, then it is really very difficult for them to use it. Because they are then into a second order of innovation, with innovation built on innovation. You do get such activity nowadays, the perception of more desirable innovation and the difficulty of producing that because you need an intermediate innovation in order to do it. But human capabilities are really very limited and you cannot really go back that far. For example, the early work in Italy on the barometer, or in England on the vacuum, required very sophisticated clear glass technology. If glass workings of a suitable quality had not been in the vicinity, there could have been no barometer or vacuum experiment. The Chinese could not have made a sophisticated barometer or vacuum chamber with rock crystal. And they certainly would not have developed glass over thousands of years just in the off-chance that it might one day be useful in order to make a barometer. It was an unrelated accident that glass was available in Italy where the barometer was also developed. People living in the Orkneys in the middle of the seventeenth century could not have made a barometer or vacuum chamber.

What has happened is that innovation is coming about in a selectionist mode by way of people using what is available to them. Yet it is more than just a matter of availability. Glass was available to the Romans, but the Romans were not potential innovators in the field of the generation of large amounts of abstract reliable knowledge. Those in mid seventeenth-century Italy or England had both the glass and the particular curiosity. So if glass had been available in the first half of the seventeenth century in China it seems very unlikely that it would have led to the discovery of the microscope, telescope and barometer. There is no reason why its presence would have had this effect. The presence of glass is a selectionist, necessary, but not sufficient, condition.

A long and wide perspective helps us to counter a view, particularly prevalent in our technologically overwhelmed and calculative age, that technologies once discovered will inevitably be retained and improved. Anthropology and archaeology have shown that the abandonment of apparently useful technologies is a widespread phenomenon through the ages. Irrigation systems have been allowed to collapse, fishing hooks abandoned, the wheel, even writing, forsaken. Thus it is no great shock to find that glass should have been more or less abandoned in eastern Asia.

The understanding of humankind must encompass our biological as well as our social evolution – hence the interactions we explored in the previous chapter in our hypothesis as to the causes of myopia.

Anthropology is famous (and infamous) for its cultural relativism. It describes and analyses different ways in which humans face the challenge of life, but on the whole it refrains from judging one as morally better than another. This is a helpful perspective when we consider the ways in which the two ends of Eurasia tried to overcome the difficulty of how to preserve and even increase reliable knowledge.

Some years ago, a Japanese economic historian and demographer, Akira Hayami, made a distinction between the civilisations of Asia, which tried to increase agricultural and craft production by increasing human labour, what he called an 'industrious revolution', and the civilisations at the other end of Eurasia which did so by replacing human labour by machines and non-human energy, what is famously called an 'industrial revolution'. In a curious way it is possible to see the same divergence of strategies in the attempt to increase the amount of reliable knowledge in the world.

In eastern Asia there was an 'industrious' revolution – the extension of literacy, the development of woodblock printing, the multiplication of written characters, the extensions of the schooling and examination systems. This put huge pressures on the human eye, which became, through progressive myopia, a sort of surrogate magnifying glass, able to do what could only be done by glass tools in the west. It was an intensification of intellectual production which has some interesting parallels to the extreme attention to detail and intensification that is found in the many craft skills practised there and in the production of wet rice.

At the western end of Eurasia, the human body was not forced in the same way. Increasingly the chief instrument of knowledge acquisition was strengthened by glass instruments, just as the human muscles were supported by new tools that made wind, water and animal power a huge supplement to human labour. So a whole set of 'machines' for thinking – glasses in old age, prisms and magnifying glasses, mirrors, telescopes and microscopes – was developed. An industrial revolution of the intellect took place as a counterpart to the industrious revolution of the east.

The relativism of the anthropologist would not place one as 'better' than the other; they were two different approaches, each of which had their strengths and weaknesses. It just happens that the path which leads through industriousness starts to reach the limits imposed by the law of diminishing returns much sooner than the path which leads through machines of thought, which is still showing potential in a world so heavily based on glass, whether in computing, optic fibre networks or television and photography.

Yet we should not push the analogy to 'industrious' and 'industrial' too far since it would then become misleading and divert us from other, more important, differences in this context. The core of the difference may lie elsewhere. In the west there was a growing understanding that there was a different class of knowledge that could be generated, experimentally, and by 'torturing' nature. As this realisation grew via a loose but effective European network, resting on institutions such as universities, originally conceived for quite other purposes, so approaches were developed to further generate knowledge, to make it more reliable and to communicate it with increasing precision. In the East, such possibilities hardly evolved, because the existence of this body of knowledge was not recognised and the curiosity to pursue it diligently was weak. In this context, the presence or absence of clear glass becomes a pivotal, enabling device. Putting it more forcefully, even if China and Japan had developed fine clear glass on a large scale, it is still doubtful that there would have been the development of reliable knowledge that occurred in parts of the west.

In all of this, of course, it is essential to remember that glass is only an enabling, perhaps necessary, but far from sufficient cause of the massive transformation in the methods of obtaining and the overall quantity of reliable knowledge. Glass, we would argue, was a necessary cause in the development of new thought systems in the west and their absence in the east, but it was not sufficient in itself. Many other things were needed. Such a complex outcome was the result of a multitude of interacting pressures. Thus this book is just one part of the story of the emergence of the modern world.

Appendix 1
Types of Glass

THREE DIFFERENT GROUPS or types of glass are discussed in this book; soda, potash and lead glass. A fourth type, the Venetian cristallo, is mentioned, though this is a form of soda glass. The major constituent of all these glasses is silica, chemically known as silicon dioxide, quartz or rock crystal. Sand is granulated quartz. Indeed, glass can be thought of as quartz, which in its pure form is very difficult to work, which has been modified by melting with other chemicals. This produces a material which can be readily shaped at a more practical temperature as it goes through its semi-liquid stage.

Silica comprises approximately 44 per cent of the earth's mantle, by far the most abundant compound. It melts at 1726 degrees centigrade, much too high a temperature for early furnaces. In addition, when it does melt, it does so quite suddenly, and does not go through a gradual softening range, as it is heated, which enables glass to be shaped so effectively.

In order to reduce the melting temperature, silica, usually in the form of washed white sand, is melted with another chemical compound such as soda (sodium carbonate) or potash (potassium carbonate). Soda melts at 851 degrees centigrade and potash at 901. This is roughly the 'white heat' which can be seen when there is a good draught on a charcoal or coke fire.

At these temperatures and above, they decompose producing large volumes of the gas carbon dioxide. This, together with air (oxygen and nitrogen) trapped in the original mix, produces a very frothy melt. Modern glass-making is done in furnaces at 1500 to 1600 degrees, at which temperature the glass is very fluid and the larger bubbles can rise to the top. In addition, small amounts of arsenic oxide or antimony oxide can be added, which assist in the removal of the very small bubbles.

These advantages were not available until the middle of the nineteenth century. Furnace temperatures were much lower and a multitude of even quite large bubbles can be seen in most museum specimens of glass.

Glasses made from silica and soda or silica and potash alone are not stable; they slowly disintegrate from the action of water, even moisture in the atmosphere. Stability is given to glass, both soda and potash glass, by the presence of a small percentage (5 to 10 per cent) of calcium oxide. This was often introduced quite accidentally, as an impurity such as broken shells, in the sand. A typical composition for soda glass would be 73 per cent silica, 17 per cent sodium oxide (from the soda), 5 per cent calcium oxide (from limestone, or accidental calciferous inclusion, such as sea shells) and 5 per cent other oxides such as magnesium or aluminium oxide.

There are large deposits of soda in some parts of the world and it was from deposits in Egypt that supplies for the eastern Mediterranean glass industry were obtained until around 700–800 AD. Supplies then became more difficult to obtain and the ash of a sea marsh plant, barilla, which contained soda, was widely used.

In many inland areas, soda was more difficult to obtain so potash was substituted for soda. Potash is contained in the burnt ash of many plants, bracken and beechwood are typical sources, and so forests that were near a source of clean sand became sites for glass production, the so-called 'forest glass'. The medieval description of glass-making by Theophilus, a German monk writing around 1120 AD, describes glass-making using wood ash.

Potash glass would typically contain 10–13 per cent and like soda glass is often accidentally stabilised by the presence of lime as an impurity in the raw materials. Even so, potash glass is more subject to disintegration or severe surface weathering due to water than soda glass. The proportions of the major constituents of glass, that is silica, soda, potash and lime, are very variable and different glass-making centres had different customs about the mixes that they used. Indeed, it is often possible nowadays to make a fairly good guess about the place of origin of a medieval glass from around Europe or the Middle East by a careful analysis of its constituents.

Sand, even the sand we call 'white sand', is usually not white at all, but a brownish colour. Most of this brown is due to iron oxides which end up in the glass, colouring it with a greenish tint. Old glass bottles can be deep green, while even modern window glass displays a green edge.

Clear glass is best made using pure raw materials; the favoured source of silica for the famous Venetian cristallo glass was the white quartz pebbles from the bed of the river Ticino, which flows down from the Swiss Alps and through northern Italy. These pebbles were roasted in a furnace and then pulverised before being mixed with the soda from barilla, which had itself been purified by dissolving in water and then recrystallising. The choice of these pure raw materials produced a clear and readily worked glass, but unfortunately the lime (calcium oxide) content was often too low, and the finished articles were rather prone to disintegration over time.

Glasses containing high percentages of lead oxide were made in China before the tenth century and in Murano, the glass-making island of the Venetian lagoon, in the early seventeenth century. Lead glass has an attractive appearance, giving a high lustre or sparkle, and is easy to cut and polish on a wheel. It was used in a coloured form for producing artificial gemstones.

In England in the 1670s a clear lead glass was developed (a typical composition was 51–60 per cent silica, 28–38 per cent lead oxide and 9–14 per cent potash), to compete with the Venetian glass that was being imported in large amounts. This new glass was highly successful, producing heavy, robust, but lustrous glasses and bowls, well suited to engraving. As a material for containers, lead glass would not occupy a significant place in our story of the rise of western civilisation, but fortuitously it had some remarkable optical properties. These were not recognised for over fifty years after its original manufacture, but they enabled substantial improvements in the quality of the images of first telescopes and then microscopes to be achieved.

Appendix 2
The Role of Glass in Twenty Experiments that Changed the World

IT WAS SIX OR SEVEN YEARS AGO that we first had a hunch that there might be something particularly interesting about the role of glass in the unfolding of our scientific-cum-industrial western civilisation. It was a further two or three years until we realised that the almost complete absence of clear glass in Asia before the seventeenth century provided a way of comparing innovation, including innovation of new knowledge, as between the two ends of Eurasia.

The first hunch that glass was absolutely crucial to western development was based on the observation that the microscope, telescope and barometer had revolutionised biology, medicine, astronomy and chemistry. In order to obtain a broader view of the importance of glass in science and hence in the development of the modern world we have analysed the twenty experiments that have been chosen by the much respected historian of science at Oxford, Rom Harré, in his book *Great Scientific Experiments: Twenty experiments that Changed Our View of the World* (Phaidon, Oxford, 1981).

Harré did not use the presence or absence of glass as a criterion in his choice of experiment. So these twenty experiments can act as a rather random and objective way in which to reinforce or refute the view that glass was of central importance in the development of science.

Of the twenty experiments which are described, sixteen could not have been performed without the use of glass apparatus, sometimes as a transparent container, sometimes as optically worked components such as prisms or lenses. Here is a brief description of the experiments and the role of glass in them. Dates are necessarily imprecise. Sometimes, as with Aristotle, exact dates are not known; sometimes because the experiment was developed over several years.

1 *Aristotle*, *c.* 350 BC, described the development of the chick embryo. Glass not used.
2 *William Beaumont*, American army doctor, *c.* 1833. Taking advantage of a permanent hole in the stomach lining of an army porter, following a musket accident, he experimented on the process of digestion. Glass containers were used to observe digestion (ceramic containers could, with considerable disadvantages, have been substituted). Temperature both inside the stomach and in external experimental jars was measured with a liquid-in-glass thermometer.
3 *Robert Norman*, *c.* 1581. The first carefully recorded measurements of magnetic dip, that is the way in which a balanced compass needle, after magnetising, will dip down towards the earth as well as seeking North and South. The compass needle was floated beneath the surface of water in a large clear wine glass.
4 *Stephen Hales*, *c.* 1716–24. The establishing of the way in which sap moves upwards and downwards in a plant. These movements were observed within a glass tube attached to the cut end of a branch via a lead pipe.

5 *Konrad Lorenz*, *c.* 1926–38. Experiments to determine the conditions of imprinting, the acquisition of fixed behavioural patterns by the young of a species during their early development. No glass required.
6 *Galileo*, *c.* 1603. Investigation into the nature of acceleration using a smooth bronze ball rolling down a groove cut into a long sloping wooden beam, and timing the motion with the pulse. No glass required.
7 *Robert Boyle and Robert Hooke*, *c.* 1662. Measurement of the volume of a quantity of air trapped in one closed leg of a u-shaped tube, under the influence of various 'heads' of mercury in the other, open-topped leg of the u-shaped tube. The ability to produce a long, strong glass tube, to bend it into shape and to seal one end was essential to this experiment.
8 *Theoderic of Freiburg*, *c.* 1300. A spherical blown glass flask, possibly a urine flask, was filled with clear water and held up above eye level and while facing away from the sun. This successfully simulated the role of raindrops in the formation of rainbows. Glass was required and used optically as a shaped container.
9 *Louis Pasteur*, 1880. The preparation of artificial vaccines to protect against infectious disease, by the creation of a series of cultures with an extended time period between each. The use of the optical microscope with its glass lenses was essential to the identification and isolation of the microbes concerned.
10 *Ernest Rutherford*, *c.* 1919. The first instance of the artificial transmutation of one element (nitrogen) into another (hydrogen). The apparatus used was largely made of glass, used as a container for the gases in the experiment.
11 *A. A. Michelson and E. W. Morley*, 1887. An attempt to

detect the movement of the earth in space by comparing the velocity of light in two different directions at right angles to each other. The results were negative, but a good negative result is often as useful as a positive one and it contributed to a Nobel Prize for Michelson. The apparatus was high precision optics using glass for many components of lenses and mirrors.

12 *Jacob and Wollman*, 1956. This experiment explores mechanisms of heredity and the direct transfer of genetic material between bacteria. The use of the optical microscope was an essential tool in the identification and isolation of the bacteria.

13 *J. J. Gibson*, 1962. Gibson was concerned with the way in which we perceive the everyday objects that surround us in our ordinary lives. He conducted experiments which showed that perception was a far more complex process than a simple matter of seeing and touching. No glass was required.

14 *Antoine Lavoisier*, 1774. Mercury was heated in air, which produced an oxide of mercury which itself was heated, thus regenerating oxygen, in a glass bell-shaped jar, the base of which was standing in a tub of water. The heating was performed over a period of twelve days by means of a large glass lens which focused rays from the sun through the glass bell jar upon the mercury. Glass in different forms was essential at each stage of the experiment.

15 *Humphry Davy*, 1808. Davy isolated potassium, sodium and six other elementary metals from melts of their salts by passing an electric current through the melt. Glass was not a strict requirement for the main experiment (ceramic parts, for instance, could have been used instead of glass for the outer containers for the batteries he used). But

glass was certainly required for the containers to collect and analyse the gases evolved and hence to determine the nature of the reaction.

16 *J. J. Thompson*, 1897–1903. In a sealed glass tube containing gas at very low pressure – the forerunner of the modern cathode-ray tube used in television sets – the deflection of a beam of cathode rays by means of electrically charged deflector plates demonstrated the existence of particles smaller than atoms. These we now call electrons.

17 *Isaac Newton*, 1672. Newton used three glass prisms and a glass lens to separate white light from the sun into a spectrum of colours, to recombine them again to white light and then to separate them again.

18 *Michael Faraday*, *c.* 1833. Electricity was known to be produced from a number of sources, the cell or battery, friction with an insulator such as amber or glass, by moving a magnet with relation to a conducting wire, by heating the junction of two different metals, and 'animal' electricity produced, for example, by the electric eel. Faraday performed a series of experiments to establish that, however produced, the electricity from each source was really identical. Glass was used extensively both as an insulator and in the construction of electricity storage devices known today as Leyden jars.

19 *J. J. Berzelius*, 1810. An extensive series of experiments which determined accurately, for the first time, the atomic weights of forty-five elements. For the precise measurement of volumes of liquids, in which he was a pioneer, and in the handling of gases, glass apparatus was essential.

20 *Otto Stern*, 1923. Stern demonstrated that beams of atoms or molecules created by the vaporisation of the materials

concerned, behave both as particles and waves. Glass is not essential to the main apparatus, but is used in the ancillary equipment, vacuum pumps and photographic plates.

Further Reading

In order to keep the text uncluttered, we have dispensed with footnotes and bibliographic citations. This book is, however, highly dependent on the work of others. The following section indicates some of the books and articles which have been used for each chapter and provides further reading for those who would like to pursue specific issues. The full titles of the works are listed in the bibliography below. If you would like further information on glass and its history please visit *www.alanmacfarlane.com/glass*.

1 Invisible Glass

The section on glass in Lewis Mumford's *Technics and Civilisation* was a major influence at the start of working on this book. On clocks and printing see Landes, Eisenstein. On the strangeness of glass as a substance see McGrath and Frost, Honey, *Encyclopedia Britannica* under 'Glass'. For some other stimulating general works which cover various aspects of the themes in this book see Adams, Crosby, Gregory, Park, Perkowitz.

2 Glass in the West – from Mesopotamia to Venice

For general studies of glass which cover most or all of this period see Battie and Cottle, *Encyclopedia of Glass*, Tait, Honey (various), Klein and Lloyd, Singer et al., *Encyclopedia*

Britannica under 'Glass' and 'Mirrors', Liefkes, McGrath and Frost, Vose, Derry and Williams, *Chambers Encyclopedia* under 'Glass', Dauma, *Encyclopedia of World Art*, vi, Hayes, Moore, Bray, Zerwick.

On early and Roman glass: Allen, James and Thorpe, Bowerstock et al. On glass between 400–1200: Hayes, Dopsch, Wilson, Theophilus. On glass from 1200–1700: Godfrey, Braudel, Mumford, Davies, Anglicus, Ashdown, Houghton, L'Art du Verre.

3 Glass and the Origin of Early Science

We have relied principally on the works of Crombie and Lindberg. Other useful background works are Huff, Needham, *The Shorter Science*, Park, Ludovici, Bernal. On magical thought and science see Kittredge, Yates, Walker.

4 Glass and the Renaissance

There is a vast literature on the Renaissance. Here are a few of the authors we found useful.

General background: Burckhardt, Hale, Larner, Burke, Gottlieb, Fry, Clark, Hay. Art: Panofsky, Gombrich, Baxandall, Wolfflin, Harbison, Hauser, Witkin, Arnheim, Miller, Friedlander, Baltrusaitis, Gell, Ayres, Bazin, Kahr. On perspective: Damisch, White, Bunim, Wright, Ivins, Kemp, Edgerton.

On the the psychology of perception: Blakemore, Gregory; contemporary accounts: Alberti, Leonardo da Vinci, Vasari. On individualism and its causes: Carrithers et al., Gurevich, Morris, Abercrombie, Macfarlane (*Origins*). On autobiography and individualism see Delany. On painting and the mirror, Mumford. On Japanese attitudes to mirrors: Benedict, 202; Koestler, 173; Riesman and Riesman,

273. For mirrors and their mysteries, Gregory.

5 Glass and Later Science

For early suggestions concerning the relations between glass and science which originally inspired this chapter, see Mumford, *Technics*. For the connection between glass and science see Singer et al., iii; Derry and Williams. On microscopes, see Ludovici, Mills. For barometers, Knowles Middleton. For the twenty scientific experiments, see Harré.

6 Glass in the East

On Islamic glass-making: Battie and Cottle, Tait, Klein and Lloyd, Honey, Singer et al., iii, 230. The best single source is Oliver Watson's article in Liefkes. Islamic architecture: Blair and Bloom, Talbot Rice, Bazin. Glass in India: Singh, Dikshit, Klein and Lloyd, Battie and Cottle, Liefkes. Glass in China: Phillips, Liefkes, Needham, Klein and Lloyd, Tait, Needham (*Shorter Science*, 4); Battie, Cottle and Temple, Elvin, 83–4; Needham, 'Optick Artists'. On mirrors, Balustraitis. On architecture in China: Williams, 726–3; Fortune, 79–90; Hommel. On painting on glass: Jourdain and Soame Jenyns, Crossman, ch. 8. Glass in Japan: Blair, Klein and Lloyd. European accounts of glass in Japan: Kaempfer, iii, 72; Thunberg, *Travels*, iv, 59, iii, 279; Oliphant, 189; Screech, 134–6; Alcock, i, 179.

7 The Clash of Civilisations

For the introduction of glass into China, see Liefkes, Battie and Cottle, Phillips. For the China trade, see Crossman, Osborne ch. 8. For Japan and the introduction of glass, see Screech. The description of art in China and Japan is based on Sullivan, Binyon, La Farge, Bowie, Clunas, Dyer Ball under 'Art'. See also Needham and Wang.

8 Spectacles and Predicaments

The effects of spectacles are discussed in Mumford, Larner, Davies, Landes (*Wealth*). See also Elvin, 83–4 and Needham, 'Optick Artists'. On myopia and eye sight: Trevor-Roper (various), Grosvenor and Goss, Dobson, Tokoro, Goodrich, Chan, Mann and Pirie, Souter, Harman, Browning, Parsons and Duke-Elder, Price, Jollily, Nielsen, Eden, Gregory, *Chambers Encyclopedia* under 'Eye Care', 'Myopia', 'Vitamins'; *Encyclopedia Britannica*, under 'Vision', Weston. There is a massive documentation of the history of ophthalmology (in seven volumes) in Hirschberg. On Japanese houses and furniture, see Morse. The title of this chapter is taken from a book on social theory by the late Ernest Gellner.

9 Visions of the World

For a useful preliminary overview of theories of the origins of the scientific revolution, see Shapin, *Scientific*. For a longer overview of theories, Cohen, *Scientific Revolution*. For the Renaissance, the seminal book by Burckhardt, *The Civilisation of the Renaissance in Italy* is still unsurpassed. A modern study is Hale, *The Civilisation of Europe in the Renaissance*. Some of the theoretical issues concerning causation and methodology are discussed in Macfarlane, *Savage*, *Riddle* and *Making*, which also discuss other aspects of the origins of the modern world. The distinction between 'industrious' and 'industrial' was originally developed by Akira Hayami of Keio University.

Sources for Quoted Passages

(Full titles to all the texts from which quotations are taken are in the bibliography below.)

1 Invisible Glass

McGrath and Frost, *Glass*, p. 297; Honey, *Glass*, p. 1; Dr Johnson (*Rambler* no. 9, 17 April 1750) is quoted in McGrath and Frost, *Glass*, p. 5.

2 Glass in the West – from Mesopotamia to Venice

Tait, *Glass*, p. 166 (Agricola); W. R. Lethaby, quoted in McGrath and Frost, *Glass*, p. 104 (medieval cathedrals). The reference to Houghton comes from the article on glass in the *Encyclopedia Britannica*.

3 Glass and the Origin of Early Science

Einstein's famous remark on the origins of science is quoted in Crombie, *Science*, p. 41 The quotations from Grosseteste and Bacon are from Crombie, *Science, Optics*, pp. 198, 201–2.

4 Glass and the Renaissance

Gombrich on picture writing, Gombrich, *Art*, p. 144; Filarete on Brunelleschi, quoted in Damisch, *Origin*, p. 63 (Italian phrases omitted); Leonardo on master of painters, *Note-*

books, vol. ii, p. 219; Leonardo on painting in a mirror, *Painting*, p. 202; Leonardo on perspective on a pane of glass, quoted in Wright, *Perspective*, p. 87; Leonardo on visual pyramids, *Notebooks*, vol. ii, p. 48; Leonardo on use of veil, *Painting*, p. 216. The quotations from Thunberg are in his *Travels*, vol. iii, p. 284.

5 Glass and Later Science
Bacon, *New Atlantis*, p. 214; the quotation from Knowles Middleton from *Barometer*, p. 3.

7 The Clash of Civilisations
All Du Halde quotations are from Du Halde, vol. ii, pp. 126–8; Williamson, quoted in Battie and Cottle, p. 214; Macartney Embassy, in Cranmer-Byng, p. 299; Thunberg, *Travels*, vol. iv, p. 60; Screech, pp. 137, 194; Sullivan, pp. 41–2; Clunas, p. 184; Sullivan, p. 43; Clunas, pp. 176–7; and Sullivan, pp. 43, 58, 80; Bowie, pp. 75–6; Rasmussen, *Spectacles*, p. 56; Gombrich, *Illusion*, p. 73; Sullivan, p. 19; Rasmussen, *Spectacles*, p. 56; Binyon, *Flight*, p. 10.

8 Spectacles and Predicaments
The first quote is from Bird, p. 165. The quotations from Rasmussen are from his pamphlet *Spectacles*. The television review was by Robert Hanks in *The Independent*, 7 July 1999. The other quotations, in order, are from the following: Bird, p. 165; Macartney Embassy – Cranmer-Byng, p. 299; Hommel, p. 197; Browning, p. 109; Dyer Ball, p. 227; Trevor-Roper, *Blunted*, p. 42; Morse, vol ii, pp. 105-6; Li Yu, p. 117 ; Mann and Pirie, pp. 146, 159, 160; Trevor-Roper, *Blunted*, pp. 21–2.

Bibliography

Unless specified otherwise place of publication is London.

Abercrombie, Nicholas, Hill, Stephen and Turner, Bryan S., *Sovereign Individuals of Capitalism* (Allen & Unwin 1986)

Adams, Robert McC., *Paths of Fire: An Anthropologist's Inquiry into Western Technology* (Princeton University Press, Princeton, 1996)

Alberti, Leon Battista, *On Painting*, trans. Cecil Grayson (Penguin 1991)

Alcock, Sir Rutherford, *The Capital of the Tycoon: A Narrative of a Three Years' Residence in Japan* (Longman 1863)

Allen, Denise, *Roman Glass in Britain* (Shire Archaeology, Princes Risborough, Bucks., 1998)

Anglicus, Bartholomaeus, *On the Properties of Things* [De Proprietatibus Rerum], trans. John Trevisa, 2 vols. (Clarendon Press, Oxford, 1975)

Arnheim, Rudolf, *Art and Visual Perception, a Psychology of the Creative Eye* (University of California Press, Berkeley, 1957)

— *Visual Thinking* (Faber & Faber 1970)

Ashdown, Charles, *History of the Worshipful Company of Glaziers of the City of London* (Blades 1919)

Bacon, Francis, *The Advancement of Learning and New Atlantis* (Oxford University Press, Oxford, 1951)
Baltrusaitis, Jurgis, *Anamorphic Art*, trans. W. J. Strachan (Cambridge University Press, Cambridge, 1977)
Battie, David and Cottle, Simon (eds.) *Sotheby's Concise Encyclopedia of Glass* (Conran 1997)
Baxandall, Michael, *Painting and Experience in Fifteenth-Century Italy*, 2nd edn (Oxford University Press, Oxford, 1988)
Bazin, Germain, *A Concise History of Art: Parts 1 & 2*, trans. Francis Scarfe (Thames & Hudson 1962)
Benedict, Ruth, *The Chrysanthemum and the Sword: Patterns of Japanese Culture* (Routledge 1967)
Berenson, Bernard, *The Italian Painters of the Renaissance* (Fontana Library 1960)
Bernal, J. D., *Science in History* (Watts 1957)
Binyon, Lawrence, *The Flight of the Dragon, an Essay on the Theory and Practice of Art in China and Japan, Based on Original Sources* (John Murray 1948)
— *Painting in the Far East, an Introduction to the History of Pictorial Art in Asia Especially in China and Japan* (Dover Publications, Mineola, NY, 1969, originally published in 1934)
Bird, Isabella, *Unbeaten Tracks in Japan* (Virago Press 1984)
Biringuccio, Vannoccio, *The Pirotechnia of Vannoccio Biringuccio: The Classic Sixteenth-Century Treatise on Metals and Metallurgy*, trans. and ed. Cyril Stanley Smith and Martha Teach Gnudi (Dover Publications, Mineola, NY, 1990)
Blair, Dorothy, 'Glass' in *Encyclopedia of Japan*, vol. 3 (Kodansha International, Tokyo, 1983)
— *A History of Glass in Japan* (Kodansha International,

Tokyo, 1973)

Blair, Sheila S. and Bloom, Jonathan M., *The Art and Architecture of Islam 1250–1800* (Yale University Press, New Haven, CT, 1994)

Blakemore, Colin, *Mechanics of the Mind*, BBC Reith Lectures 1976 (Cambridge University Press, Cambridge, 1977)

Bowerstock, G. W. et al. (eds.), *Late Antiquity. A Guide to the Postclassical World* (Harvard University Press, Cambridge, MA, 1999)

Bowie, Henry P., *On the Laws of Japanese Painting. An Introduction to the Study of the Art of Japan* (Dover Publications, Mineola, NY, 1952, originally published in 1911)

Braudel, Fernand, *Civilisation and Capitalism 15th–18th Century*, trans. from French, 3 vols. (vol. i: *The Structures of Everyday Life*) (Collins 1981, 1983, 1984)

Bray, Charles, *Dictionary of Glass: Materials and Techniques* (A. & C. Black 1995)

Browning, John, *Our Eyes, and How to Preserve Them From Infancy to Old Age, with Special Information about Spectacles* (Chatto & Windus 1896)

Bunim, Miriam Schild, *Space in Medieval Painting and the Forerunners of Perspective* (AMS Press, New York, 1970)

Burckhardt, Jacob, *The Civilisation of the Renaissance in Italy* (Phaidon Press 1960)

Carrithers, Michael, Collins, Steven and Lukes, Steven (eds.), *The Category of the Person, Anthropology, Philosophy, History* (Cambridge University Press, Cambridge, 1985)

Chambers Encyclopedia, New Revised Edition, 1966

Chan, Eugene, 'The General Development of Chinese

Ophthalmology from its Beginnings to the 18th Century', *Documenta Ophthalmologica*, 68 (1988), 177–84

Clark, Kenneth, *Civilisation* (BBC 1971)

Clunas, Craig, *Pictures and Visuality in Early Modern China* (Reaktion Books 1997)

Cohen, H. Floris, *The Scientific Revolution: A Historiographical Inquiry* (University of Chicago Press, Chicago, 1994)

Cranmer-Byng, J. L. (ed.), *An Embassy to China: Being the Journal Kept by Lord Macartney During his Embassy to the Emperor Ch'ien-lung 1793–4* (1962)

Crombie, A. C., *Augustine to Galileo*, 2 vols. (Mercury Books 1964)

— *Robert Grosseteste and the Origins of Experimental Science 1100–1700* (Clarendon Press, Oxford, 1953)

— *Science, Art and Nature in Medieval and Modern Thought* (Hambledon Press 1996)

— *Science in the Middle Ages*, vol. i, *Medieval and Early Modern Science* (Doubleday 1959)

— *Science, Optics and Music in Medieval and Early Modern Thought* (Hambledon Press 1990)

Crosby, Alfred, *The Measure of Reality, Quantification in Western Society 1250–1600* (Cambridge University Press, Cambridge, 1998)

Crossman, Carl L., *The Decorative Arts of the China Trade, Paintings, Furnishings and Exotic Curiosities* (Antique Collectors' Club 1997)

Damisch, Hubert, *The Origin of Perspective*, trans. John Goodman (MIT Press, Cambridge, MA, 2000)

Dauman, Maurice (ed.), *A History of Technology and Invention. Progress Through the Ages*, vol. ii, *The First*

Stages of Mechanization 1450–1725, trans. Eileen B. Hennessy (John Murray 1980)
Davies, Norman, *A History of Europe* (Oxford University Press, Oxford, 1996 under 'Murano')
Delany, Paul, *British Autobiography in the Seventeenth Century* (Routledge and Kegan Paul 1969)
Derry, T. K. and Williams, Trevor I., *A Short History of Technology, from Earliest Times to AD 1900* (Clarendon Press, Oxford, 1960)
Dikshit, M. G., *History of Indian Glass* (University of Bombay, Bombay, 1969)
Dobson, Roger, 'The Future is Blurred', *The Independent*, 20 May 1999
Dopsch, Alfons, *The Economic and Social Foundations of European Civilisation* (RKP 1953)
Dyer Ball, J., *Things Chinese*, 5th edn (Singapore, 1925)
Eden, John, *The Eye Book* (Penguin 1981)
Edgerton, Samuel Y., Jr, *The Renaissance Rediscovery of Linear Space* (Harper & Row 1975)
Eisenstein, Elizabeth L., *The Printing Press as an Agent of Change: Communications and Cultural Transformations in Early-Modern Europe* (vols. i and ii, complete in one volume, Cambridge University Press, Cambridge, 1980)
Elgin Mission – *see* Oliphant
Elvin, Mark, *Another History, Essays on China from a European Perspective* (Sydney University 1996)
Encyclopedia Britannica, 11th edn (Cambridge Press, Sydney, 1910)
Encyclopedia of World Art, vol. vi (McGraw Hill, New York, 1962)
Fortune, Robert, *Three Years' Wanderings in the Northern Provinces of China* (John Murray, 1847, facsimile edn,

Time-Life Books, Alexandria, VA, 1982)

Friedlander, Max J., *From Van Eyck to Bruegel, Early Netherlandish Painting* (Phaidon 1956)

Fry, Roger, *Vision and Design* (Penguin 1940)

Gell, Alfred, *Art and Agency, An Anthropological Theory* (Clarendon Press, Oxford, 1998)

— 'The Technology of Enchantment and the Enchantment of Technology', in Jeremy Coote and Anthony Shotton (eds), *Anthropology, Arts and Aesthetics* (Clarendon Press, Oxford, 1992)

Godfrey, Eleanor S., *The Development of English Glassmaking 1560–1640* (Clarendon Press, Oxford, 1975)

Gombrich, E. H., *Art and Illusion: A Study in the Psychology of Pictorial Representation* (Phaidon Press 1962)

— *The Image and the Eye: Further Studies in the Psychology of Pictorial Representation* (Phaidon Press 1999)

— *Meditations on a Hobby Horse, and Other Essays on the Theory of Art* (Phaidon 1963)

— *The Story of Art* (Phaidon Press 1950)

Goodrich, Janet, *Perfect Sight the Natural Way: How to Improve and Strengthen your Child's Eyesight* (Souvenir Press 1996)

Gottlieb, Carla, *The Window in Art* (Abaris Books, New York, 1981)

Gregory, R. L., *The Intelligent Eye* (World University 1971)

Gregory, Richard, *Mirrors in Mind* (W. H. Freeman, Oxford, 1997)

Gregory, Richard, Harris, John, Heard, Priscilla and Rose, David (eds.), *The Artful Eye* (Oxford University Press, Oxford, 1995)

Gurevich, Aaron, *The Origins of European Individualism*, trans. Katharine Judelson (Blackwell, Oxford, 1995)
Hale, John, *The Civilisation of Europe in the Renaissance* (HarperCollins 1993)
Harbison, Craig, *The Art of the Northern Renaissance* (Everyman Art Library 1995)
Harman, N. Bishop, *The Eyes of our Children* (Methuen 1916)
Harré, Rom, *Great Scientific Experiments* (Phaidon Press 1983)
Hauser, Arnold, *The Social History of Art*, 3 vols. (Vintage Books, New York, 1957)
Hay, Denys (ed.), *The Age of the Renaissance* (Guild Publications 1986)
Hayes, E. Barrington, *Glass through the Ages* (Penguin 1959)
Henkes, H. E., 'History of Ophthalmology', *Documenta Ophthalmologica*, 68 (1988), 177–84
Hirschberg, Julius, *History of Opthamology* 7 vols (1982–6).
Hockney, David, *Secret Knowledge* (Thames & Hudson, 2001).
Hommel, Rudolf P. *China at Work: An Illustrated Record of the Primitive Industries of China's Masses, whose Life is Toil, and Thus an Account of Chinese Civilisation* (MIT Press, Cambridge, MA, 1969)
Honey, W. B., *English Glass* (Bracken Books 1987)
— *Glass: A Handbook for the Study of Glass Vessels of all Periods and Countries and a Guide to the Museum Collection* (Victoria and Albert Museum, London, published by the Ministry of Education 1946)
Huff, Toby, E., *The Rise of Early Modern Science, Islam, China, and the West* (Cambridge University Press, Cambridge, 1993)

Ivins, William M., *Art and Geometry: A Study in Space Intuitions* (Dover, Mineola, NY, 1964)

— *On the Rationalization of Sight, with the Examination of Three Renaissance Texts on Perspective* (Da Capo Paperbacks, New York, 1975)

Jackson, C. R. S., *The Eye in General Practice*, 7th edn (Churchill Livingstone 1975)

James, Peter and Thorpe, Nick, *Ancient Inventions* (Michael O'Mara Books 1995)

Jourdain, Margaret and Soame Jenyns, R., *Chinese Export Art in the Eighteenth Century* (Scribner, New York, 1950)

Kaempfer, Englebert, *The History of Japan, Together with a Description of the Kingdom of Siam 1690–1692*, trans. J. G. Scheuchzer, 3 vols. (1906, facsimile edition, Curzon Press, Richmond, 1993)

Kahr, Madlyn Millner, *Velasquez. The Art of Painting* (Harper and Row, New York, 1976)

Kemp, Martin, *The Science of Art, Optical Themes in Western Art from Brunelleschi to Seurat* (Yale University Press, New Haven, CT, 1990)

Kittredge, George Lyman, *Witchcraft in Old and New England* (Russell & Russell, New York, 1956)

Klein, Dan and Lloyd, Ward (eds.), *The History of Glass* (Black Cat 1992)

Knowles Middleton, W. E., *The History of the Barometer* (Baros Books 1994)

Koestler, Arthur, *The Lotus and the Robot* (Hutchinson 1960)

La Farge, John, *An Artist's Letters from Japan* (Hippocrene Books, 1986)

Landes, David S., *Revolution in Time. Clocks and the Making of the Modern World* (Harvard University Press, Cambridge, MA, 1983)

— *The Wealth and Poverty of Nations, Why Some Are So Rich and Some So Poor* (Little, Brown & Co. 1998)

Larner, John, *Culture and Tradition in Italy 1290–1420* (Batsford 1971)

L'Art du Verre section of *L'Encyclopedie de Diderot et D'Alembert* (eighteenth century, reprinted in facsimile by Inter-Livres, no date)

Liefkes, Reino (ed.), *Glass* (Victoria & Albert Museum 1997)

Lindberg, David C., *The Beginnings of Western Science. The European Scientific Tradition in Philosophical, Religious, and Institutional Context, 600 BC to AD 1450* (University of Chicago Press, Chicago, 1992)

— *Roger Bacon and the Origins of Perspectiva in the Middle Ages* (Clarendon Press, Oxford, 1996)

— *Roger Bacon's Philosophy of Nature. A Critical Edition* (Clarendon Press, Oxford, 1983)

— *Theories of Vision from Al-Kindi to Kepler* (University of Chicago Press, Chicago, 1976)

— (ed.), *John Pecham and the Science of Optics, Perspectiva communis* (Wisconsin University Press 1970)

Ludovici, L. J., *Seeing Near and Seeing Far* (John Baker 1966)

Macfarlane, Alan, *The Making of the Modern World: Visions from West and East* (Palgrave 2002)

— *The Origins of English Individualism* (Blackwell, Oxford, 1978)

— *The Riddle of the Modern World: Of Liberty, Wealth and Equality* (Macmillan 2000)

— *The Savage Wars of Peace, England, Japan and the Malthusian Trap* (Blackwell, Oxford, 1997)

Mann, Ida and Pirie, Antoinette, *The Science of Seeing* (Penguin 1946)

McGrath, Raymond and Frost, A. C., *Glass in Architecture and Decoration* (Architectural Press 1961)

Miller, Jonathan, *On Reflection* (National Gallery Publications 1998)

Mills, A. A., 'Single-Lens Magnifiers', parts i–vi, *Bulletin of the Scientific Instrument Society*, nos. 54–9 (1997–8)

Moore, N. Hudson, *Old Glass. European and American* (Tudor Publications, New York, 1935)

Morris, Colin, *The Discovery of the Individual 1050–1200* (University of Toronto Press, Toronto, 1995)

Morse, Edward S., *Japanese Homes and Their Surroundings* (first edition, 1886, Dover Publications, Mineola, NY, 1961)

Mumford, Lewis, *The Myth of the Machine: The Pentagon of Power* (Harcourt Brace, New York, 1970)

— *Technics and Civilisation* (George Routledge 1947)

Needham, Joseph, 'The Optick Artists of Chiangsu' with Lu Gwei-Djen, in Jerome Ch'en and Nicholas Tarling (eds.), *Studies in the Social History of China* (Cambridge University Press, Cambridge, 1970)

— *The Shorter Science and Civilisation in China*, 4 vols. abridged by Colin A. Ronan (Cambridge University Press, Cambridge, 1980(2), 1995, 1994)

— (ed.) *Science and Civilisation in China*, vols ii, iii, iv:1, 2, v:2, 3 (Cambridge University Press, Cambridge)

Nielsen, Harald, *Medicaments Used in the Treatment of Eye Diseases in Egypt, the Countries of the Near East, India and China in Antiquity* (Odense University Press, no date)

Oliphant, Laurence, *Narrative of the Earl of Elgin's Mission to China and Japan in the Years 1857, '58, '59*, vol. ii only (Blackwood, Oxford, 1859)

Osborne, Harold (ed.), *The Oxford Companion to the Decorative Arts* (Oxford University Press, Oxford, 1975)
Panofsky, Erwin, *Early Netherlandish Painting, its Origins and Character*, 2 vols. (Icon Editions, New York, 1971)
— *The Life and Art of Albrecht Dürer* (Princeton University Press, Princeton, NJ, 1971)
Park, David, *The Fire within the Eye, a Historical Essay on the Nature and Meaning of Light* (Princeton University Press, Princeton, NJ, 1997)
Parsons, J. H. and Duke-Elder, S., *Diseases of the Eye*, 11th edn (Churchill 1948)
Perkowitz, Sydney, *Empire of Light, a History of Discovery in Science and Art* (Joseph Henry Press, Washington, DC, 1996)
Phillips, Phoebe (ed.), *The Encyclopedia of Glass* (Spring Books 1987)
Popper, Karl R., *Conjectures and Refutations: The Growth of Scientific Knowledge* (Routledge and Kegan Paul 1978)
Price, Weston A., *Nutrition and Physical Degeneration* (published by the author, 1945)
Rammazzini, Bernardo, *A Treatise on the Diseases of Tradesmen* (1705)
Rasmussen, O. D., *Chinese Eyesight and Spectacles* (Tonbridge Free Press). A copy of this rare pamphlet, revised in 1950, is in the Needham Centre Library, Cambridge.
— *Old Chinese Spectacles*, 2nd edn, revised (Northchina Press, 1915). There is a copy of this rare pamphlet in the Needham Centre Library, Cambridge.
— *A Thesis on the Cause of Myopia*. A copy of this rare booklet of 1949 is in the Cambridge University Library, classmark 9300.c.913.
Riesman, David and Riesman, Evelyn Thompson

Conversations in Japan, Modernization, Politics and Culture (Allen Lane 1967)

Screech, Timon, *The Western Scientific Gaze and Popular Imagery in Later Edo Japan, The Lens Within the Heart* (Cambridge University Press, Cambridge, 1996)

Shapin, Steven, *The Scientific Revolution* (University of Chicago Press, Chicago, 1996)

Singer, Charles, Holmyard, E. J., Hall, A. R. and Williams, T. I. (eds.), *A History of Technology* vols. ii–v (Clarendon Press, Oxford, 1972)

Singh, Ravindra N., *Ancient Indian Glass: Archaeology and Technology* (Parimal Publications, Delhi, 1989)

Souter, W. N., *The Refractive and Motor Mechanism of the Eye* (Keystone Publications, Philadelphia, 1910)

Sullivan, Michael, *The Meeting of Eastern and Western Art* (University of California Press, Berkeley, 1989)

Tait, Hugh, *The Golden Age of Venetian Glass* (British Museum 1997)

—(ed.), *Five Thousand Years of Glass* (British Museum 1991)

Talbot Rice, David, *Islamic Art* (Thames & Hudson 1975)

Temple, Robert, *The Genius of China, 3,000 Years of Science, Discovery and Invention*, Introduction by Joseph Needham (Prion 1991)

Theophilus, *On Divers Arts. The Foremost Medieval Treatise on Painting, Glassmaking and Metalwork*, trans. John G. Hawthorne and C. S. Smith (Dover Publications, NY, 1979)

Tokoro, Takashi, 'Vision Care in Japan', *The Vision Care*, pp. 47–52 (Proceedings of the Yoya Vision Care Conference, April 1998)

Trevor-Roper, Patrick D., *Lecture Notes on Ophthalmology* (Blackwell Science, Oxford, 1961)

—'The treatment of myopia', (*British Medical Journal* vol. 287, 17 December 1983, pp. 1822–3)

—*The World Through Blunted Sight: An Inquiry into the Influence of Defective Vision on Art and Character* (Allen Lane 1988)

Vasari, Giorgio, *The Lives of the Artists*, selection trans. George Bull (Penguin 1965)

Vinci, Leonardo da, *The Notebooks of Leonardo da Vinci*, arranged and trans. by Edward MacCurdy, 2 vols. (Cape 1938)

— *The Notebooks*, 2 vols., ed. Edward MacCurdy (Reprint Society 1954)

— *On Painting*, ed., Martin Kemp (Yale University Press, New Haven, CT, 1989)

Vose, Ruth Hurst, *Glass* (Collins Archaeology, Collins 1980)

Wecker, John, *Eighteen Books of the Secrets of Art and Nature* (London, S. Miller, 1660)

Weston, H. C., *Sight, Light and Work*, 2nd edn (H. K. Lewis, London 1962)

White, John, *The Birth and Rebirth of Pictorial Space* (Faber 1957)

Williams, S. Wells, *The Middle Kingdom: A Survey of the Geography, Government, Literature, Social Life, Arts and History of the Chinese Empire and its Inhabitants* 2 vols. (W. H. Allen 1883)

Wilson, David, *The Anglo-Saxons* (Pelican 1975)

Witkin, Robert W., *Art and Social Structure* (Polity Press, Cambridge, 1995)

Wolfflin, Heinrich, *Principles of Art History: The Problem of the Development of Style in Later Art*, trans. M. D. Hottinger (G. Bell 1932)

Wright, Lawrence, *Perspective in Perspective* (Routledge and Kegan Paul 1983)

Yates, Frances A., *Giordano Bruno and the Hermetic Tradition* (Routledge and Kegan Paul 1964)

Yu, Li, *The Carnal Prayer Mat* (Wordsworth reprint 1995)

Zerwick, Chloe, *A Short History of Glass* (Harry N. Abrams, New York, in association with The Corning Museum of Glass, 1990)

Index

Figures in italics refer to captions.

F

G

O

P